DE CHARDIN

UNDERSTANDING PIERRE TEILHARD DE CHARDIN

A Guide to
THE PHENOMENON OF MAN

by
MAURICE KEATING
with
H. R. F. KEATING

LONDON
LUTTERWORTH PRESS

First published 1969

LONDON

LUTTERWORTH PRESS, 4 BOUVERIE STREET, E.C.4

7188 1601 3

*Printed in Great Britain by
Richard Clay (The Chaucer Press) Ltd,
Bungay, Suffolk*

CONTENTS

ACKNOWLEDGMENTS

IN THIS ATTEMPT to simplify much of Teilhard's *The Phenomenon of Man* the authors have deliberately restricted themselves to the field of natural phenomena dealt with in this book, and have not included comment on Teilhard's Christology or any other spiritual aspect of his writings.

For the Prologue they have drawn on some of the information contained in Claude Cuenot's *Biographical Study* published in 1965 by Burns & Oates, and for further biographical details would refer readers to Robert Speaight's *Biography* published by Collins in 1967. For permission to draw freely from the text of *The Phenomenon of Man*, acknowledgment is also made to Messrs. Collins, who first published the work in English in 1959.

PROLOGUE

PIERRE TEILHARD DE CHARDIN was born in the French province of Auvergnes in 1881. At the age when other children experienced their first feelings of attraction to persons, art or religion, because of his extraordinary attraction to "little bits of metal", he would withdraw secretly from company to contemplate and perhaps even worship them.

At that tender age he could see with extraordinary clarity a whole series of idols made of iron. In the country he kept a plough key hidden carefully away in the corner of a yard; in the home, he made his own a metal staple which protruded from his nursery floor. Later, little shell splinters which he collected lovingly at a near-by shooting range became objects of his veneration. This created in him a sense of dedication, and in later years he accepted the fact that his whole spiritual life had as its base the development of this unusual expression of worship.

As Sir Julian Huxley records in his prologue to *The Phenomenon of Man*, at the age of ten Teilhard went to a Jesuit college where he was able to develop his fascination with iron into systematic field geology and mineralogy. At eighteen he decided to become a Jesuit. During the course of his studies, part of which were at the Jesuit College, then at Hastings, Sussex, he acquired a considerable competence in palaeontology, the study of fossils, which was to become the key to his philosophy. Back in Paris, Teilhard's interests were first directed to what was to become the centre of his life's work—the evolution of man.

But the even tenor of his life was soon interrupted by the

1914–18 War, during which period he served as a corporal stretcher-bearer, receiving the Military Medal.

He was invited to work as a palaeontologist in China in 1922, where he was destined to stay for twenty years with occasional excursions to the U.S.A., Abyssinia, India, Burma and Java.

This was the period when he developed what was first intended as a thesis on evolution and eventually became *The Phenomenon of Man*.

In 1933 Teilhard began to have trouble with his superiors concerning his views and he was instructed to refuse any situation or official appointment that might be offered him in Paris. Then in 1944 he was refused ecclesiastical permission to publish *The Phenomenon of Man* which he had submitted to Rome. Other rebuffs followed, but he was to obtain some solace in being appointed an officer of the Legion of Honour. Nevertheless opposition to his philosophy did not diminish and in 1951 he felt obliged to remove to New York to work. There he continued his researches, writings and development in a milieu he found sympathetic until his death in 1955.

The present volume makes no attempt to cover the details of his remarkable work in the field of palaeontology, but rather it attempts to re-express in as simple a manner as possible the most significant of his works, *The Phenomenon of Man*, first published in English by Collins in 1959, originally in French in 1955.

After reading this book one reader said, "Nothing can ever be viewed in quite the same light again. It is a classic in its own right." It deals with the origins of matter on earth, of atoms and the gradual evolution of life, of the animal origins of man which finally lead to "cellular intelligence" and the idea of the over-lordship on earth of man himself.

During his lifetime Teilhard de Chardin was able to convince his immediate followers of the rightness of his fundamental conviction that science and religion are, if not synonymous, at least two conjoined aspects of life. Though first of all a scientist, he had an inspired belief in the presence of a Universal God. In particular he claimed that life on earth is part of a grand evolutionary process which moves relentlessly and irreversibly towards the greater integration of the human race, leading eventually to the welding together of all mankind.

It has been said that by his life and work, Teilhard threw a bridge across the centuries-old chasm between science and religion. That is precisely what he has done, and at the same time he has given man his due place in the Universe.

The Phenomenon of Man is difficult reading, and this attempted simplification is primarily intended to provide the reader with an introduction to a remarkable conception of Man and his Universe.

Teilhard, more perhaps than any thinker in history, has been able to break the barrier between the concrete world of the scientists and the inspired visions of the giants of religion through the ages.

I

HOW TEILHARD SAW THE UNIVERSE

EVERYTHING MUST HAVE a beginning. Where better to
start than with the smallest thing we know—the atom?

Atoms have a nucleus around which electrons spin at a
fantastic speed. They really consist of three essential factors,
for besides their nucleus and their electrons there is the energy
activating these which, by reacting upon each other, form a
whole, the intrinsic power of which is almost impossible to
penetrate, or split. The fact that it has been split with such
devastating effect is common knowledge, but until recently
this idea was regarded as quite impossible and unbelievable.

One of the first things that Teilhard draws our attention to
is the way in which the world divides itself into ever smaller
particles. The process, he says, is easily visible in our every-
day experience, in raindrops, in grains of sand, in the multitude of the stars, in the hosts of the living and even in the
ashes of the dead. Man needs neither microscope nor elec-
tronic equipment in order to realize that he is surrounded by
and resting on dust. But to count the grains and describe
them requires all the patient craft of modern science. Every
particle of matter under the physicists' instruments tends to
reduce itself to ever smaller and smaller proportions until
we near the infinitely small—where nothing is visible at
all except as impulses interpreted by the electron microscope.

All our normal associations of light, colour, warmth, etc., lose their meaning here. Seen from this point of view, the world we thought we could see and touch appears to be only something in a continuous state of disintegration.

But, by contrast, the more we split and pulverize matter the more its fundamental unity becomes apparent. There emerges an astonishing similarity in the elements so examined. It is almost as if the stuff, the material, of which we are made, of which everything is made, proves to be reducible in the end to one simple kind of substance.

We find that however small the "heart" of an atom is, its influence extends, potentially at least, over every other atom. What is it that holds atoms together? What is the mysterious force that cements them? The name given by science to this all pervading principle is simply energy.

Energy is the measure of what passes from one atom to another in the course of the transformations they undergo. It is the unifying power between them and represents the most primitive form of universal stuff. This energy is the only really indivisible thing: an all-pervading force that binds the whole together.

Thus, the entire system of the universe holds together. If you try to cut out a portion of this network, you fail because you cannot do so without its becoming unravelled at the edges. So only one way of considering the universe is really possible: by taking it as a whole, in one piece.

Yet between this structure of energy and the structure of matter there is, paradoxically, nothing in common. Matter in its various orders of magnitude does not repeat its different combinations time and time again, as energy does. Matter starts with a simple series of nuclei and circling electrons. But it then goes on to more complex extensions of simple bodies. Next it changes to combinations of molecules, inexhaustible in number. And, to jump from the infinitesimal to the infinite, we come at the end to the universe of stars and galaxies.

In these multiple zones of the universe, order and design appear only in the whole. The actual stuff of the universe, though woven in the simple mesh of energy, does not, when it becomes matter, repeat itself from one point to another. But it still represents a single whole.

This Whole Teilhard defines as Space. In it the radius of action of each atom—that is, its potential sphere of influence —extends to the utmost limits. Seen in this light, the atom is no longer the microscopic closed world we may have imagined it to be. It is the infinitesimal centre of the universe itself.

If we now turn our attention to the total of all those infinitesimal centres, we find that, indefinite though their number may be, they constitute together a growth which has definite effects, among them this earth as we know it. But our world appears to us as it does, only when we also see it in terms of time or duration.

Teilhard employs a vivid image here. Space, he says, is no more than a section, cut at any one instant, of a vast tree whose roots plunge down into the abyss of the unfathomable past and whose branches rise up to a future that, at first sight, has no limit. We may add that such a section of this tree, cut so thin as to have no duration, would naturally be impossible for us to know; it is only when the section has some thickness that we realize the tree exists.

In this perspective, Teilhard says, "the world appears like a mass in process of transformation". Seen against time, the stuff of the universe is becoming concentrated into ever more organized, more complex, forms of matter. It goes from the level of electrons, to that of atoms, from atoms to

3

molecules, from simple molecular combinations to vastly more complicated ones.

But where do these changes take place—starting, say, with the framework of the molecules? It is not haphazardly at any point in space. It is in the heart and surface of the stars. Surely, Teilhard says, a "chemistry" of the heavens must fill out the story of the atoms? It seems, he says, that the making of more complex forms of material can take place only under cover of a previous concentration of the stuff of the universe in the stars. "The stars are laboratories in which the evolution of matter proceeds in the direction of large molecules." And, he adds, this happens according to definite rules.

These rules, he says, are the two scientific laws of the conservation and the dissipation of energy. The first states that during physical or chemical changes no new energy is developed; energy simply passes from one constituent to another. The second law is that in every physico-chemical change a fraction of the available energy is irrecoverably lost in the form of heat or light.

Under these rules the transformations which constitute the evolution of matter appear as a costly operation in which an original impetus slowly becomes exhausted. A wearing-away seems to be gradually consuming the cosmos in its entirety. The scientific picture of the universe is, then, of an eddy rising on the surface of an inevitably descending current.

Teilhard comments:

So says science: and I believe in science: but up to now has science ever troubled to look at the world other than from *without*?

THE WITHIN OF THINGS

TEILHARD HOLDS THAT the quarrel between the hard-and-fast scientist and the believer in an inner interpretation of things arises largely from their inability to find any common ground, rather than from any inherent impossibility of reconciling two apparently contradictory points of view.

He sees the materialist scientists as always talking about things as though they consisted only of observable actions in transient relationships; and, on the other hand, the believers in inner interpretation he sees as obstinately determined to regard all things as being the mere manifestations of a spiritual force. Both think on different planes; they never meet; each sees only half the question.

He is convinced, however, that the two sides could join in sharing a point of view which would take into account both the external aspect of the world and the internal aspect of things. Any other course, he believes, cannot deal with every single phenomenon that the world presents to the human consciousness.

Having described the Without of things, i.e. that which can be measured and interrelated, he goes on to offer a description of the Within of things. He argues out three statements:

That there is in fact a Within.

That this Within has some connections with the developments which science recognizes, and is capable of being expressed in terms of quality.

That this Within has some connections with science's developments that can be expressed in terms of quantity.

First, Teilhard illustrates his conception of everything possessing a Within by pointing out how to a physicist nothing exists except what is measurable, the Without of things, and that this intellectual attitude will still serve the bacteriologist, who can treat the cultures he grows in his laboratory as things which react on each other in measurable ways. But then, Teilhard points out, the botanist, moving on the realm of plant life, begins to encounter obvious difficulties in keeping strictly to terms within which everything can be dealt with in figures. And when you come to the insectologist and the zoologist the difficulties become progressively greater and greater. Finally, with the study of man it becomes impossible not to take into account the unmeasurable activities of the mind because we know them to exist from our own experience.

Teilhard goes on to argue that in fact the Within, which we cannot escape taking into account in the case of man, exists all the way down the scale as well. It is simply that in observing from our limited viewpoint we fail to see it. He cites the parallel case of our observations of matter to reinforce the point. By normal standards of human existence, he says, the mountains and stars are models of majestic changelessness. But, observed over a sufficiently great length of time, we now know that the earth's crust changes ceaselessly under our feet, that mountains move.

The restriction of consciousness to the higher forms of life, Teilhard argues, is equally merely a question of appearances deceiving. It is not sufficient to regard the fact that a human being is capable of thought as being a mere abnormality, an aberration from the norm and therefore not a proper subject for scientific investigation. Other abnormal activities are investigated by scientists. The apparently abnormal behaviour of radium is a telling instance. This scientists were prepared to probe, and were thus led to the advance of modern physics.

The difference between the activity of radium and that of consciousness is that radium's effects can be measured, whereas to integrate consciousness into a systematic scheme of things means taking into consideration the existence of a new aspect or dimension in the universe.

But, Teilhard points out, it is the task of research workers to discover the universal hidden beneath the exceptional.

An irregularity in nature, he says, is only the sharp exacerbation of something diffused throughout the universe in a state which eludes our detection. Provided a phenomenon is properly observed, even if it occurs only in one spot, it is of necessity present, though concealed, everywhere else.

It is not enough to say "Consciousness is completely evident only in man and is therefore an isolated instance of no interest to science." What must be said is: "Consciousness is evident in man and therefore, though half seen in this one flash of light, it extends everywhere."

This conclusion, he says, is pregnant with consequences and is, by sound analogy with all the rest of science, inescapable.

Consciousness, taken in its widest sense to include everything from the most rudimentary forms of interior perception imaginable to the phenomenon of reflective thought in human beings, can equally be called the Within.

Teilhard cites in confirmation of these views a passage from *The Inequality of Man*[1] by J. B. S. Haldane, the physiologist and geneticist. It runs:

[1] Chatto, 1932, p. 113.

We do not find obvious evidence of life or mind in so-called inert matter, and we naturally study them most easily where they are completely manifested; but if the scientific point of view is correct, we shall ultimately find them, at least in rudimentary forms, all through the universe. Now, if the co-operation of some thousands of millions of cells in our brain can produce our consciousness, the idea becomes vastly more plausible that the co-operation of humanity, or some sections of it, may determine . . . a Great Being.

Teilhard himself returns to this theme at a later stage in his argument.

But at this point we are concerned with developing our consideration of the properties of the Within. Teilhard first postulated its existence and then put forward the concept that the Within has some connection with the developments which science recognizes, capable of being expressed in terms of quality. How can we, in short, define the rules by which the Within, for the most part entirely hidden, suddenly shows itself?

Teilhard suggests a way in which we can do this by presenting three observations which do not take on their true value till they are linked together.

First Observation: That the Within of things, like the Without, takes an atomic, or particle, form.

Second Observation: That consciousness traced back along the path of evolution becomes ever less complicated, less differentiated, less well defined.

Third Observation: That material complexity and a perfection of consciousness go hand in hand.

Having made these observations, Teilhard asks us to picture the universe as moving from one state, in which it had a very large number of very simple material elements, to another in which it has a smaller number of very complex groupings.

In the first state, centres of consciousness, because they are extremely numerous and at the same time extremely loosely connected, reveal themselves only by overall effects which are subject to the laws of statistics. That is, taken collectively they obey the laws of mathematics. They are thus the proper field of the physicist.

In the second state, on the other hand, the less numerous and more highly individualized elements gradually escape from the slavery of large numbers. Or, as Teilhard puts it, they allow their basic spontaneity to break through and reveal itself. We can begin to see them one by one, and in so doing we enter the world of biology.

Turning to consider the connections the Within has with science's developments as expressed in terms of quantity, Teilhard remarks that, though the concept of spiritual energy is absolutely familiar to us, such energy is so intangible that the whole universe can be described in mechanical terms without taking it into account at all. He says he is unable to put forward a truly satisfactory solution to the problem of the link between spiritual and physical energy. But he does offer one example of a line of research that could be adopted and the kind of interpretation that might be followed.

We cannot but be aware, he says, that mind and body act on one another. The mind is dependent on the body. "To think, we must eat." The loftiest speculation, the most burning love are paid for by an expenditure of physical energy. Sometimes to inspire thought we need bread, sometimes wine, sometimes even the action of a drug or a hormone, sometimes the stimulation of a colour, sometimes the magic of a sound.

Without the slightest doubt, Teilhard goes on, there is

something through which material and spiritual energy hold together and are complementary. In the last analysis, somehow or other there must be a single, fundamental energy operating in the world.

Yet as soon as we try to couple mind and matter together their mutual independence becomes as clear as their mutual relation.

> To think, we must eat. But what a variety of thoughts we get out of one slice of bread!

The line of solution Teilhard offers starts by making the assumption that essentially all energy is quasi-mental in character. Then in each particular element he divides it into two separate components, one the force, "tangential energy", linking the element with all others of the same complexity as itself in the universe, and the other the force, "radial energy", drawing the element forward towards an ever greater complexity and centricity and ever more directed towards the future.

The element, supposing that it disposes of a certain free tangential energy, must be in a position to increase its internal complexity in association with neighbouring elements, and thereupon to augment its radial energy. The latter will then be able to react in its turn in the form of a new composition in the tangential field. And so on.

Teilhard goes on to enlarge on how these two forms of energy have operated from the time when thousands of millions of years ago a fragment of matter composed of particularly stable atoms was detached from the surface of the sun and began to condense, roll itself up and take shape as a planet containing within its globe and orbit the future of man.

3

THE EARTH IN ITS EARLY STAGES

WHEN THE EARTH was first detached from the sun it was still enveloped by the gases of the stratosphere and the atmosphere. Its centre contained the metals of the barysphere. But between these two extremes lay a series of complex substances originating in the stellar system. There were minerals and carbonates in the lithosphere, and water and other liquids in the hydrosphere.

From the outset energy was liberated. Essential oxides such as the minerals, water and carbon dioxide were formed because of the affinities of their elements. In rocks there occurred perpetual transformation of the mineral species.

This however was a limited process. By their innate structure the molecules of minerals are unfitted for further growth.

But this was not the only field of advance. Added to the energy liberated in the crystallizing process was that released by the atomic decomposition of radioactive substances and that given off by the sun's rays. This energy began building up certain carbonates, hydrates and nitrates. In these, particles grouped themselves and exchanged positions in a theoretically endless network. Molecules here could link with molecules to form ever larger and more complex molecules.

This world of organic compounds is ours. We live among them and are made of them.

But so accustomed have we become to the idea that we are part of this world that we have got into the habit of associating it only with life as already constituted.

Whereas, if we wish to pinpoint the place of man in nature it is essential, Teilhard says, to restore the organic compounds of which man is composed to their rightful place in the hierarchy of the universe.

It is Teilhard's reiterated belief that nothing can ever burst forth across any threshold of evolution that has not already existed in an obscure and primordial fashion. Unless the compound organic factor had existed on earth from the first moment, he says, it could never have sprung into being later.

It was in these organic compounds that the Within of the earth, already discussed, was gradually to be concentrated.

This Within, the non-material aspect of that portion of the stuff of the universe enclosed from the beginning of time within the narrow confines of the earth, must at every point be related to the Without, to the material aspect.

Teilhard says that a certain mass of elementary consciousness must have originally been imprisoned in the matter of the earth. The world already carried pre-life inside it. And, moreover, it did so in a definite quantity.[1]

To show how from this primitive quantum all the rest emerged, Teilhard says, it is enough to compare stage by stage the general laws, referred to earlier, for the development of radial energy with the development of the earth outlined above. Radial energy, he says, by its very nature increases in radial value in step with the increasing chemical complexity of the elements of which it represents "the inner

[1] This philosophical interpretation has been attacked, particularly since Teilhard himself, in a postscript to *The Phenomenon of Man* modifies his previous statement to comment that matter and spirit do not present themselves as "things" or "natures" but as simple related variables.

lining". But the elements are increasing in complexity in conformity with the laws of thermodynamics. Put the two propositions side by side and they shed light on one another. They tell us that pre-life was no sooner enclosed in the embryonic earth than it emerged from the torpor to which it appeared to have been condemned by its diffusion in space. Its activities were set in motion step by step with the awakening of the forces of synthesis enclosed in matter.

The earth was possibly born by accident when it became detached from the sun. But scarcely had this accident occurred than it was immediately turned to good purpose. The envelope round the mass of the earth, where it was possible for elements to attain greater and greater complexity, was an organic whole by virtue of the fact that its elements could unite to form these more complex structures. This unity was reflected in the Within. It formed the basis for increases in intensity from age to age until eventually something burst out upon the early earth. And that was Life.

4

THE ADVENT OF LIFE

No EXACT POINT can be fixed for the advent of life, Teilhard says. It has always been there in the form of pre-life, since everything has always existed in some extremely attenuated form. This does not however alter the fact that there was a moment when life as we know it burst into being. When anything exceeds a certain stage of development it suddenly changes its condition. It was in this way that pre-life became life.

An immense period of time after the earth was formed, it became cool enough to allow the formation on its surface of chains of molecules. At this time it was probably covered by a single great sea, from which emerged the first signs of the continents.

An observer would probably have seen only an inanimate liquid desert. The waters contained no trace of mobile particles even seen through a microscope. He would have observed only inert aggregates.

But, after a sufficient lapse of time, those same waters must here and there have begun writhing with minute, microscopic things. And from that initial proliferation there has stemmed the amazing profusion of living things whose matted complexity came to form what Teilhard calls the biosphere, the layer of life.

One thing is certain, Teilhard says. This change cannot

14

have been the result of a simple continuous process. There must have a been a threshold, a coming to maturity. By analogy with all we have learnt from the comparative study of natural developments, this particular moment must have represented the beginning of a new order.

He then goes on to set out first what must have been the nature of this transformation, and then the limits in space and in time in which it occurred. And to precede this he attempts an explanation of the event that fits in with both what are believed to be the conditions prevailing in the earliest ages of the earth and with those of the earth today.

He begins by looking at the origin of life from an external, material point of view. In this light, he says, it begins with the cell. Here lies the secret of the connection between the worlds of physics and biology. The cell is, he says, the natural granule of life in the same way as the atom is the natural granule of matter.

Hitherto, he goes on, scientists have regarded the cell from the point of view of the higher forms of life it gives rise to. But, like the moon in its first quarter, the cell has been illumined only on the side that looks towards its future evolution, leaving the other side (the layers he has called pre-life) floating in darkness.

But, he says, the significance of the cell cannot be understood unless it is situated on an evolutionary line between a past and a future. We must focus on its origins if we want to grasp the essence of its novelty. We have become too accustomed to think of the cell as an object without antecedents. What happens if we look on it at one and the same time as the outcome of a long preparation and yet as something totally original? As a thing that is born?

Taking what appear to be the simplest cells in nature, we find, he says, the unmistakable characteristics of the granular formations, the simplicity, the structural symmetry, the infinitesimal size. We are still on the first rung of life.

And the advances in biochemistry of the first half of the twentieth century, Teilhard continues, are beginning to reduce to measurable proportions the gap between protoplasm and mineral matter.

He points out that the zoological separation of living creatures into different species, as made in the eighteenth century by Lamarck, actually revealed and measured a difference in evolutionary age. In other words, the time factor enters into the classification of different animal species. Thus he is also able to equate the difference in the development of molecules and cells with different eras in time. In the discovery of viruses, we have for example to allocate another interval of time in the life of the universe, and introduce a hitherto forgotten era (an era of sub-life) in the immense series of ages that measure the past of our planet.

Without a long period of maturing no profound change can take place in nature. In Teilhard's view, for example, an evolutionary break of the first order must have taken place with the appearance of the first cells.

The Cellular Revolution

In what way can the nature of this break from pre-life into life be envisaged? Teilhard asks. Externally, he regards the cell as an organism of extraordinary complexity and yet of a no less extraordinary fixity in its fundamental type. He describes it as a triumph of multiplicity organically contained within a minimum of space. He doubts if it can be compared to anything either in the world of "animates" or that of "inanimates". We are right to look on these as the first of living forms, but equally entitled to view them as another state of matter—something as original in its way as the electron, the atom or the crystal. In other words, a new type of material for the new stage of the universe—part of the stuff of the universe re-appearing with all its characteristics, but reaching a higher rung of complexity and advanced still further in consciousness.

Teilhard looks at the within of the cell in another way. Having assumed that "conscious" life was present in the world before the first appearance of the cell, he goes on to explain in what specific way the internal (radial) energy of the cellular unit is modified to correspond with its new external (tangential) constitution.

Having already endowed the long chain of atoms and molecules with a rudimentary free activity, it is not by a totally new beginning, but by a major constitutional change that the cellular revolution expresses itself. He considers that the realization of this cellular awakening prepares the way for the advent of the phenomenon of man himself.

He justifies this by explaining that any essential change-over between two states or forms of consciousness can and does happen at other levels of evolution. The degree of Withinness of a cosmic element can undoubtedly vary to the point at which it rises suddenly to another level. The critical change in the intimate arrangement of the elements induces *ipso facto* a change of nature in the state of consciousness of the particles of the universe.

For confirmation of this he asks us to look at the astounding spectacle displayed in the transit to life on the surface of the early earth; at the thrust forward in spontaneity; at the luxuriant unleashings of fanciful creations; at the unbridled expansion and the leaps into the improbable.

Before considering the evolutionary consequences of this transit, Teilhard asks his readers to look a little more closely into the conditions of its historical realization. He admits that it is difficult to do this by referring only to the intuition of our senses and suggests a more indirect method of trying to picture "new-born life". He does this by stages.

Stage 1. The Milieu

Teilhard asks us to go back with him perhaps a thousand million years and wipe out of our minds most of those material superstructures which form the features of the earth's surface today. He pictures our planet as enveloped in a shoreless ocean through which, at a few isolated points, protuberances of future continents had begun to emerge by seismic or volcanic eruption. It was in this heavy, liquid and active ocean that the first cells must have been formed. The best we can do is to imagine them in terms of granules of protoplasm. But if their structure is incomprehensible we do know that they were incredibly small and that there must have been a bewildering number of them.

Stage 2. Smallness and Number

Teilhard asks at this point for one of those efforts *to see*, about which he says that, for man to discover man and take his own measure, the acquiring of a whole series of "senses" has been necessary:

A sense of immensity of space in the placing of the orbits of the objects round us at their true distances, both for the great, like the stars, and the small, like the atom;

A sense of depth, a pushing backwards through endless series and vast distances of time, a process which a sort of sluggishness of mind tends continually to reverse, squeezing all the past into one thin layer;

A sense of number, the discovering of the bewildering multitude of elements involved in the slightest change in the universe;

A sense of proportion, the realization as best we can of the difference in physical scale which separates the atom from the nebula, the infinitesimal from the immense;

A sense of quality, or of novelty, distinguishing in nature certain absolute stages of perfection and growth, without upsetting the physical unity of the world;

A sense of movement, perceiving the irresistible developments hidden in extreme slowness, or seeing the extreme agitation concealed beneath a veil of immobility—the entirely new insinuating itself into the heart of the monotonous repetition of the same things;

A sense of the organic, discovering physical links and structural unity in what appear to be mere successive states or random collections of things.

He goes on to say that our eyes may be familiar with the field of vision of a microscope without our realizing the vast difference which separates the world of mankind from that of a drop of water. Yet with the help of a microscope we can measure creatures in hundredths of a millimetre without even attempting to transplant them mentally from their scale to our own. This we must do if we are ever to probe the secrets of nascent or granular life. When, under a microscope, we lose sight of bacteria, they are no more than 1/5000th of a millimetre long. Cells, too, are not only microscopic, but literally innumerable.

There seems to be in the universe, he says, a natural relationship between size and number. Either because small creatures are faced with a relatively greater space, or else to compensate for their reduced radius of effective action, the smaller they are the more they swarm.

Stage 3. The Origin of Number

As we get near the threshold of life, Teilhard asserts that

though in the first instance cells occurred "at a single point or a small number of points, the first cells multiplied almost instantaneously—as crystallization spreads in a super-saturated solution. For surely the early earth was in a state of biological super-tension".

What, he asks next, is the most suitable way of imagining the beginnings of such a mass of living organisms? The nascent cellular world shows itself to be already infinitely complex. The phenomenon of life can be understood only as an organic problem of masses in movement, and not just a statistical problem of large numbers.

Stage 4. Inter-relationship and Shape

Once the dimensions and spontaneity of a cell are reached, a more complicated pattern appears in the stuff of the universe. Firstly, the initial mass of the cells must from the start have been inwardly subjected to a sort of interdependence which went beyond a mere mechanical adjustment, and was already the beginning of life-in-common.

The first veil of organized matter that spread over the earth could neither have established nor maintained itself without some network of influences and exchanges which made it a biologically cohesive whole—a sort of super-organism, and, to a certain extent, not a foam of lives but a living film.

Secondly, the chemical uniformity of protoplasm at varying points has been taken as proof that all existing organisms descend from a single ancestral group. But the living world gives the same appearance of being only a part of what it might have been. Teilhard points out that taken as a whole the world of organic life represents only a single branch within and above other less progressive or less fortunate growths from pre-life. He has no doubt that other growths of a different nature also branch out, even below the level of life.

Having taken his reader this far stage by stage, Teilhard sums up:

> Seen from afar, elementary life looks like a variegated multitude of microscopic elements, a multitude great enough to envelop the earth, yet at the same time sufficiently interrelated and selected to form a structural whole of genetic solidarity.

The Season of Life

Having discussed the breakthrough into life in terms of space, Teilhard goes on to consider it in terms of time. He believes that, beside the group of phenomena subject to direct observation, there is for science a particular class of facts to be considered—those which depend neither on direct observation nor on experiment but which can be brought to light only by the exploration of the past.

The length of time separating the historical origins of two successive species, or sub-kingdoms, as Teilhard describes them, is much greater than the age of mankind. So it is not astonishing that we should live in the illusion that nothing happens any more. Matter seems dead. But, he asks, could not the next pulsation be slowly developing around us?

Teilhard points out that there is a fundamental similarity between all organic beings. This manifests itself in the absolute and universal uniformity of the basic cellular pattern— especially in animals, because of the identical solutions found for the various problems of perception, nutrition and reproduction. Everywhere too, we find vascular and nervous systems, everywhere some form of blood, everywhere eyes. Naturalists are becoming more and more convinced that the "genesis of life on earth belongs to the category of absolutely unique events that, once having happened are never repeated".

Too much is heard of ebb and flow and not enough of the process whereby the complete evolution of the planet can be defined as something both continuous and irreversible; an ever ascending curve, the points of transformation of which are never repeated; a constantly rising tide below the rhythmic tides of the ages. It is on this essential curve that Teilhard believes the phenomenon of life must be situated. From this point of view the Cellular Revolution would then be seen as a critical single point, an unparalleled moment on the curve of evolution.

Such a view has the advantage of providing a reason for the deep organic likeness which stamps all living creatures from bacteria to mankind. It also explains why we never at any point find the formation of the least living thing which is not there as a result of germination. This view, in Teilhard's opinion, has two notable consequences for science.

First, by separating the phenomena of life from the numerous other periodical and secondary events on earth, and by making it one of the principal landmarks of the evolution of the globe, it puts to rights our senses of proportion and of values and hence renews our perspective of the world.

Second, by the very fact of showing that the origin of organized bodies is linked with a chemical transformation unprecedented and unrepeated in the history of the world, this hypothesis inclines us to think of the energy contained in the living layer of our planet as developing from and within a part of a closed "quantum". Life was born and propagates itself on earth in a single impulse.

5

THE EXPANSION OF LIFE

UNDER THIS HEADING, Teilhard proposes to study a simplified but structural representation of life evolving on earth under three main headings. First, the elemental movements of life; second, the spontaneous ramification of the living mass; and third, the tree of life.

He does so in the first place from Without and only starts probing into the Within of things after he has expressed these three aspects in their elemental form. He explains that at the base of the entire process, whereby the "envelope of the biosphere" spreads its web over the face of the earth, stands the mechanism of reproduction—which is typical of life. Sooner or later each cell divides and gives birth to another cell similar to itself. There is first the original cell as a single centre and then two.

Everything in the subsequent development of life stems from this remarkable occurrence. This cell division seems to be due both to the simple need of the living particle to remedy its molecular fragility and to ensure its continued growth. The process is one of rejuvenation and of shedding. (Atoms on the other hand have an almost indefinite longevity and with it, an equivalent rigidity.) The cell must split in two in order to continue its existence.

Once introduced into the stuff of the universe, this principle of the duplication of living particles knows no limits other than that of the quantity of matter provided. In its ability to double itself and to go on doubling itself without let or hindrance, life possesses a force of invincible expansion. In contrast the increase in volume in so-called inert matter soon reaches a point of equilibrium. But no such limit appears to be set to the expansion of living substance.

The more cellular division spreads, the more it gains in force. Nothing from within or without can check this spontaneous process. By a mechanism which is the opposite of chemical disintegration it multiplies without breaking up. But there comes a point when it closes in on itself, as it were, and becomes coagulated—and then by the act of reproduction it regains the faculty for inner re-adjustment and consequently takes on a new appearance and direction.

This is a process of pluralization in form as well as in number. The living unit is at once a centre of irresistible multiplication and an equally irresistible focus of diversification.

At this stage there enters what Teilhard describes as the wonderful process of "conjugation" or fusion of two elements by which there comes into play the endless permutations and combinations of distinctive types of organisms. The grouping of living particles into complex organisms is an almost inevitable consequence of their multiplication. It was this purely mechanical necessity or opportunity to get together that brought about in the long run a definite method of biological improvement. At the bottom we find the simple aggregate, as in bacteria and the lower fungi. One stage higher comes the colony of attached cells, as with the higher vegetable forms. Higher still, by a further transformation, autonomous central cells are established controlling, as it were, organized groups of living particles.

Still later, we reach in the animal kingdom groups of animals, which can approximately be called self-organized societies. Finally, there comes the last and highest forms of aggregations—not only able to organize themselves, but capable of self-reflection—namely the vast collection of human groupings. But that must wait until further on in this study of *The Phenomenon of Man*.

Reproduction, conjugation, association ... these three activities of the cell in themselves only lead to a surface deployment of the organisms. If it had been left to their resources alone, life would have spread and varied, but always on the same level. It would have been like an aeroplane which can taxi but not become airborne. It would never have taken off.

It is at this point that a new phenomenon intervenes and acts as an upward force. There seems to be no lack of examples in the course of biological evolution, of simple transformations acting horizontally by pure crossing of characters. But when we look deeper and more generally we see that the rejuvenations made possible by each reproduction achieve something more than mere substitution. They add one to the other and their sum increases in a predetermined direction. We get diversification, the growing specialization of factors forming a single genealogical sequence—in other words, the appearance of the line as a natural unit distinct from the individual. This law of "controlled complication" is known to biologists as "orthogenesis".[1] The world conceals deep and real springs of cosmic content.

Without orthogenesis life would have only spread; with it, there is an ascent of life that is invincible. Teilhard

[1] In a footnote Teilhard points out that some biologists would like to suppress the word "orthogenesis" but he considers it indispensable for affirming the manifest property of living matter to form a system in which "terms *succeed each other* experimentally, following constantly increasing degrees of 'centro-complexity'".

explains this by picturing the process in three phases—profusion, ingenuity and indifference.

First, comes *profusion*, germs jostling each other in their millions, shoving and devouring one another, fighting for elbow room and for the best and largest living space. But nevertheless there is a great deal of biological efficiency in this struggle for life. Survival of the fittest by natural selection is not a meaningless expression, provided it is not taken to imply either a final ideal or a final explanation. By this reckless self-production life takes its precautions against mishap. It increases the individual unit's chances of survival and at the same time multiplies its chances of progress. This process, strangely, combines the blind fantasy of large numbers with the precise orientation of a specific target. It could be described as directed chance.

Next comes *ingenuity*. This Teilhard describes as the constructive facet of the upward force. He says that to accumulate characters in stable and coherent aggregates, life has to be very clever indeed. Not only has it to invent the machine, but to design it so that it is simple and resilient. It by no means follows that the sum of the parts is the same as the whole, or that, in the whole, some specially new values may not emerge. He regards it as nature's skill in continuing new entities—a triumph of ingenuity on the part of life.

Lastly, there is *indifference*. Teilhard coins the phrase "Life is more real than lives", and explains that living particles, caught up in an aggregate greater than themselves become part of a chain. "From being a centre, they become an intermediary—a link no longer just existing but transmitting."

Teilhard then adds a fourth factor which embraces all three of these phases namely, *global unity*. Taken in its totality the living substance spreads over the earth—from the very first stages of its evolution—marking out the characteristics of "one single and gigantic organism". As we see it under our very eyes today, Teilhard writes, the "front" of advancing life is neither chaotic nor continuous. It is an aggregate of fragments at one and the same time divergent and arranged in tiers—in classes, orders, families, genera and species. What we see is the whole scale of groups whose variety, order of size and relationships our modern systematic biology tries to express in names.

Teilhard again reduces to three the number of factors which contribute to this ramification of life; (*a*) Aggregates of growth; (*b*) Branching-out of growth; and (*c*) Effects of distance.

(*a*)   *Aggregates of growth*

By aggregate of growth he means the fact that, for instance, the fibres of a living mass in the process of diversification tend to draw together following a restricted number of dominant directions. At the beginning this concentration of forms round a few selected axes is indistinct and indefinite; it involves a mere increase, in certain sectors, of the number or density of the lines. Then gradually the movement takes shape. Having reached a certain degree of mutual cohesion, the lines isolate themselves in a closed sheaf that can no longer be penetrated by neighbouring sheaves. From then on, their association, the bundle, will evolve on its own, autonomously. The species has become individualized. What Teilhard calls the "phylum" has been born. Many observers still refuse to see the reality of this strand of life in the process of evolution—i.e. this phylum—this living bundle, or line of lines.

The phylum, Teilhard insists, is, first of all, a living reality. It is also elastic and "polymorphous" (i.e. many-shaped). It can be as small as a single species or as vast as a "sub-king-dom". We must be able to recognize it on every scale of dimension.

It has a dynamic nature and only comes properly into view in movement. When immobilized in time it loses its features. But looked at in proper magnification and light, it can be seen to be a perfectly defined structural reality.

"What defines the phylum, in the first place, is its 'initial angle of divergence', that is to say the particular direction in which it groups itself and evolves as it separates off from neighbouring forms.

"What defines it, in the second place, is its 'initial section'." Just as it is impossible for a chemical reaction to take place unless a certain quantity of matter is present, "the phylum cannot establish itself biologically unless, from the start, it has gathered up in itself a sufficient number and variety of potentialities". Failure to break away at a sufficient angle or the lack of a certain modicum of consistency and richness "is enough to prevent a new branch from attaining indivi-duality".

"Lastly what serves not only to define the phylum but also to classify it without ambiguity as one of the *natural units* of the world is 'its power and singular law of autonomous de-velopment'."

(b) Branching-out of growth

If we say that a phylum behaves like a "living thing", this is no mere figure of speech; in its own way it grows and flourishes.

The development of a phylum is strangely parallel to the successive stages undergone by an invention made by men— the same story as with all modern inventions, from the bi-cycle to the aeroplane, from photography to the cinema, radio and television. At the outset the phylum corresponds to the discovery, by groping, of a new type of organism that is both viable and advantageous. But this new type will not attain its most economical or efficient form all at once. For a certain time it devotes all its strength, so to speak, to groping about within itself. Try-out follows try-out, without any form being finally adopted. Then at last perfection comes within sight and from that moment the rhythm of change slows down.

Stronger now than its less perfected neighbours, the newly-born group spreads and at the same time consolidates. It multiplies, but without further diversification. It has now entered its fully-grown period and at the same time its period of stability. The flourishing of the phylum by simple dilatation (expansion) or by the thickening of the initial stalk is nevertheless a procedure which is never completely realized—except in the case of a branch that has reached the limits of its evolutionary power.

A certain number of variants are still admissible. This explains why, as it grows, the phylum tends to split up into secondary phyla, each being a variant of the fundamental type. It sub-divides qualitatively at the same time as it spreads quantitatively. The mechanism tends to come into action again, in a more attenuated form, inside each ray or radiation.

Theoretically, there is no end to this process, but in fact the process quickly begins to peter out and the final spreading of the branches goes on without any further splitting up. The final picture generally presented by a phylum in full bloom is that of a radiating centre of consolidated forms.

And, as a last touch to the whole process we find, at the heart of this radiating centre, a profound inclination towards socialization. On this subject Teilhard has already referred to the vital power of association as exemplified in ants, in bees and mankind, and suggests that they exemplify one of the most essential laws of organized motion. The fact is that once they have attained their definitive form at the end of each ray, the elements of a phylum tend to come together and form societies, just as surely as the atoms of a solid body tend to crystallize.

(c) Effects of distance

By the very rhythm of its development each line of life follows a process of contraction and expansion. It takes on the appearance of a series of knots and bulges strung like beads, a sequence of narrow clusters and spreading leaves. For the process to be seen as it really is, we should require a witness on earth simultaneously present through the whole of duration, and the very idea is absurd. In reality, the ascent of life can be understood by us only from the standpoint of a short instant, through an immense layer of lapsed time.

What is granted to our experience is thus not the evolutionary movement in itself; it is this movement corrected according to its alteration by the effects of distance. The alteration shows itself quite simply through the accentuation (rapidly increasing with the difference) of the fan-structure deriving from the "phyletic" radiations of life.

This happens in two different ways, first by intensification of the apparent dispersion of the phyla, and subsequently by the apparent suppression of the clusters or stalks. This first optical illusion is due to the ageing and the decimation of the living branches. Only an infinitesimal number of the organisms that have grown successively on the tree of life exist for us to inspect today and despite all the efforts of palaeontology many extinct forms will remain unknown to us for ever. As a result of this destruction many gaps are continually forming in the ramifications of the animal and vegetable kingdom, and the further back we go the larger the gaps are.

At this point Teilhard pertinently remarks that, since the times of Lamarck and Darwin, a favourite argument employed against those who believe in the transformation of species is their incapacity to prove the birth of a species in terms of material traces. Teilhard calls these critics the "fixed type school".

Even apart from the continued accumulation of palaeontological evidence (which Teilhard himself has to offer) there is a more weighty answer (proof in fact) with which to rebut this school. They ask to be shown evidence of such transformation in the shape of specimens of the "peduncle" or bud of the originating phylum. Their request in Teilhard's opinion is both pointless and unreasonable—since to meet it would necessitate changes in the very nature of the world and the conditions under which we perceive it today. Nothing is more delicate and fugitive by its very nature, than a beginning. As long as a zoological group is "young", its characteristics remain indeterminate. It is composed of relatively few individual units, and these change rapidly. In space as in duration this bud of a living branch amounts to a minimum of differentiation, expansion and resistance. Beginnings have an irritating but essential fragility, and the effect of time on such weak and fragile organisms may well destroy all vestiges of them.

It is the same in every domain when anything really new begins to germinate around us, we cannot distinguish it, for the simple reason that it could only be recognized by what it is going to be. When it has reached full growth we look back to locate its starting point to find that it is by now hidden from our view, destroyed or forgotten.

Embryos, peduncles and all early stages of growth fade

and vanish as they recede into the past. They are prolonged into the present only by their survivors or their fossils.

What remains of such relics, when we look back, seems to have come into existence ready made. Geologists in some future time finding a fossilized motor car or aeroplane would get the same impression as we today might get from the now extinct pterodactyl. The discovery of an advanced model only might well lead them to think that such a machine had been made in this form at the first attempt.

In order to justify these comments, Teilhard proposes to illustrate them by his study of the destructiveness of the past superimposed upon the constructiveness of the future, and thus enable us to distinguish and outline the ramifications of the tree of life—to see it in its concrete reality and to measure it.

Teilhard has now carried us through the various stages of life's developments through the atoms, molecules, cells and finally on to the "phyla".

These latter he has already described first and foremost as having reached the stage in evolution of a "collective reality".

6

THE TREE OF LIFE

TEILHARD HAS SO far concerned himself with processes in the domain of living things, originating over two hundred million years ago. In order to get a clear view of the many lines of his subsequent tree of life he suggests that it is essential to make the effort to see a period not too close, where the leaves would get in the way, and not too distant, where the branches would lack detail. He suggests the period of the great family of mammals. Because of its more recent origin, he considers mankind is still in a stage of growth, whereas mammals in general have reached maturity.

Teilhard takes first a comparatively recent branch of the mammals, the placentals, those which carry the embryo in the womb right up to maturity. He points out that this group divides into herbivores (including the grain-eating rodents), insectivores, carnivores and omnivores. Each of these dominant branches again splits up.

But, among all the variety of creatures radiating as sub-branches of the four main branches, we can also distinguish, Teilhard points out, the remains of other systems. They are the attempts to abandon life on the ground and take to the air, the water, or even, in the case of moles, to the earth.

Similarly, corresponding to the great branch of the placentals, we have the remains of another branch, the marsupials,

33

animals which have to carry their young in pouches. The flowering of this branch can be seen by an accident of geography in Australia with its abundant variety of such animals, herbivorous, carnivorous and insectivorous, of all shapes and sizes.

Having looked at the ends of the branches represented by all the mammals both placental and marsupial, Teilhard invites us to track back towards their origins. The palaeontologists have unearthed the development of the mammalian branch right down to the pre-Jurassic period, more than a hundred and fifty million years ago. But, as it disappears then, it is seen to be surrounded by another form of life, the dinosaurs, pterosaurs and ichthyosaurs. And it is clear that these represent in their variety another ramifying system, and furthermore it is a system near the end of its flowering.

And again this system can be traced back till it too disappears. But it disappears amid a yet older system, the reptiles of the Permian era, perhaps two hundred and fifty million years ago. These creatures, standing squarely on strongly articulated limbs, were probably the first quadrupeds to be established on terra firma.

Beneath this system Teilhard discerns another layer, that of amphibians, crawling over the slime that then constituted the land masses of the world, a throng of squat or serpentine creatures, the animal group emerging from the waters.

At this beginning of their sub-aerial life, Teilhard points out, the vertebrates display a surprising characteristic: in every variety the skeleton form is the same. All have four legs, and legs constructed on the same pattern. In the front pair there is always a single upper bone, then the two bones of the forearm and the five digits of the hand.

The conclusion is plain. Despite the extraordinary

variety of species land animals can only represent one single
stem of life.

Teilhard then goes on to point to the origin of this stem,
fish with limb-like fins. Once widespread, as their fossilized
remains indicate, they are still to be found in one species, the
lung-fish.

Tracing the tree yet further back through the world of the
fishes, Teilhard arrives at the earliest fishes known to science.
He points out that these have a peculiar characteristic: they
are abnormally scaly as if they were attempting to compensate for their totally cartilaginous skeletons. Go further back,
he says, and you have creatures which cannot have left any
remains as fossils as they were altogether without bones.

Having traced the tree back from the mammals to this
most primitive, totally soft form, Teilhard fills in his picture
by placing in it the remaining branches of life. Alongside
the vertebrate branch he puts the insects and the plants. The
three branches today, he says in a characteristic aside, are
locked in a struggle for the world's available space.

Teilhard next emphasizes the vast numbers of the various
species. They amount to many, many millions. He gives a
diagram showing some thirty broad divisions of the Tree of
Life including many of the species already mentioned. In it,
for example, the whole branch of mammals represents a
minor off-shoot among a teeming multitude of rival types,
of whose existence we have been largely unaware. Every
one of these multitudinous branches represents a world of its
own, in some ways as important as our own. Quantitatively
—Teilhard points out—we are only one world among many,
and the latest comers at that.

It is admittedly difficult to put these various details of life
into their time perspective, but they can safely be reckoned
in many millions of years. Teilhard compares them to an

"anatomically and physiologically coherent system of over-lapping fans". Small fans of sub-species and races, larger ones of species and genera; still larger ones of animal life. The whole assemblage, animal and vegetable, he describes as forming, by association, one single gigantic stem rooted in the depth of the overall molecular world.

From top to bottom, from the biggest to the smallest, it is one same visible structure whose design is accentuated and prolonged by the quasi-spontaneous arrangements of the unforeseen elements brought forth from day to day. Each newly discovered form finds its natural place—each is born and each grows. It is difficult to reconcile this persistent growth with the determinism of the molecules, the blind play of chromosomes, the apparent incapacity to transmit acquired characteristics by propagation. Like all things in a universe in which time is definitely established as a fourth dimension, life is, and can only be, a reality of evolutionary nature and dimension. Physically and historically it determines the position of every living thing in space, in duration and in form.

7

THE RISE OF CONSCIOUSNESS

TEILHARD NOTES THAT it is abundantly clear that organic matter is in a state of continual change, but asks what measure can be found to assess the relative value of the changes. Can we say, for example, that a mammal or even a man is more advanced, more perfect, than a bee or a rose?

He believes he can see a direction and a line of progress for life which is so well marked that its reality will be universally admitted by the science of tomorrow.

Teilhard sees a dual function at work—the ever increasing complication of "numbers", i.e. creatures acquiring more and more organs with increased sensibility; and an increased simplicity or specialization of purpose.

In this connection he goes back to what he has already said about the Within and Without of things. The essence of the real is represented by the Within of things, and the "growing" or "passing" stage, or, as he termed it earlier, the "tangential" aspect, by the Without. We have only to look into ourselves to perceive directly the interaction of the Within and the Without. We have every reason to think too that such interaction also exists in animals, more or less in proportion to the development of their brains. At the stage of evolution of the dinosaurs, the brain was absurdly small compared to the size of the body. But, as soon as we reach the mammalian stage, there is a remarkable change. The average brain becomes more voluminous and convoluted —or folded together. This is broadly true of the mammals as a whole, which, as they advance in the Tree of Life, have larger and larger brains. The general order of degree in animal form depends on the extent of this brain-growth, a feature of the tree of life which conforms to the classification of systematic biology.

This differentiation of nervous tissue stands out as a significant transformation which points to a directed advance as applied to evolution as a whole.

First then, Teilhard claims that among living creatures, the brain is the sign and measure of consciousness and, secondly, that it is continually perfecting itself with time.

At this point Teilhard reverts to the consideration of the inner or radial progress of life's energies. It is by consideration of this aspect that the development of life in the general history of our planet becomes simple to follow.

The primordial emergence of organized matter marked a critical point on the curve of evolution, he says. After this first emergence, this moving mass is carried ever onward towards more consciousness. The mammals succeeded the reptiles, the reptiles succeeded the amphibians, just as the Alps replaced the Cimmerian Mountains which in their turn replaced the Hercynian range.

Life moves constantly forward in a spiral through one zoological layer to another with something always carried over, and, at the heart of life, explaining its progression, we see the impetus of a rise of consciousness.

The impetus of the world glimpsed in the great drive of consciousness can have its ultimate source only in some inner principle, which alone could explain its irreversible advance towards higher consciousness. This vital phenomenon seems both natural and possible once the reality of a fundamental impetus has been accepted.

Teilhard gives us a concrete example of this by pointing out that, according to current thought, an animal develops its carnivorous instincts because its teeth and its claws become sharp over generations. But should we not turn this proposition the other way round? he asks. If the tiger's fangs get longer and its claws sharper, is it not rather because, following its line of descent, it receives, develops and hands on the essence of its carnivorous nature?

Teilhard accepts the Darwinian theory of the struggle for survival and natural selection. As living beings, he says, we feel within ourselves the benefits of being jolted out of our natural laziness and the rut of habit by a prod from outside. What would we do without our enemies? he asks.

This passage recalls the effect his own sudden enforced departure to work in China had on the development of his theories. Life seems to take advantage of these chance events to select them through its inner working.

In sum, Teilhard claims that evolution results from an inward tendency rather than a mechanical adaptation to environment.

To write the true natural history of the world, he says, we should be able to follow it from Within. It represents a growth process which proceeds further and further along its original line and undergoes profound readjustments at given moments. It follows the law already referred to in relation to the birth of life. Sooner or later it reaches a critical point involving some change of state.

What Teilhard claims to have discovered is an evolutionary process which he describes as an irreversible increase not only in quantity, but also in quality, i.e. of brains and therefore of consciousness.

Teilhard examines the various stages of the Tree of Life

for examples of this tendency and concludes that it is impractical in the case of the nervous system to attempt to follow the evolution of consciousness in the vegetable kingdom.

He does mention, however, the fact that certain plants trap insects, and from this deduces that the rise of consciousness is incipient even in this kingdom. In the case of insects, he points out that especially in the higher insects, a concentration of nerve ganglions in the head goes with an extraordinary wealth and precision of behaviour. Nevertheless, he concludes that the world of insects has, through the geological ages, to some extent been marking time. An insect cannot grow beyond an inch or two without becoming dangerously fragile, and Teilhard points out that the superior levels of consciousness demand physically big brains.

He also comments on the prodigious arrangements which group together, in a single living machine, the swarming hive or ant-hill. He describes this as a paroxysm of consciousness which spreads outwards from within, and becomes materialized in rigid arrangements: the exact opposite of a concentration. He therefore moves on to the mammals and primates (i.e. man, monkeys and lemurs) to illustrate the development of brain-development. If a furry quadruped seems so animated compared with an ant, so genuinely alive, it is not because of our kinship with it, Teilhard says, but because its instincts are not narrowly canalized as in the spider or the bee in a single function. Unlike the insect, the mammal has an aura of freedom, a glimmer of personality.

In this connection, Teilhard points out that there is a large group of mammals to choose from, but many of them, such as the heavy-tusked animals, must be ruled out despite their psychic vitality since they are a dead-end biologically. This leaves man and the monkeys as the main branch of developing mammal groups. Rapidly Teilhard sketches for us the appropriate branches of the Tree of life—the 32-toothed monkeys of the Old World, a parallel branch of 36-toothed South American monkeys, the lemurs, branching off lower down, and yet earlier the insectivorous species. From the Old World monkey springs the branch of the anthropoids, tailless, biggest and most alert of all. Anatomically, he points out, there is an unexpectedly slight degree of differentiation in their bones while the cranial capacity is relatively bigger than any other mammal. While other branches developed their teeth or their limbs, this ramification altered little over thousands of years. This might appear a lack of progress, but in fact it gave the anthropoids a new freedom. Specialization paralyses. Palaeontology is littered with its catastrophes. But the anthropoids were able to lift themselves through successive upthrusts to the very frontiers of intelligence. What makes the primates so interesting and important to biology is that they represent a branch of direct brain-development. In contrast, the four-legged animals, like the horse and the tiger, have to some extent become prisoners of their swift-moving or predatory ways.

In the case of the primates, on the other hand, evolution, neglecting everything else, went straight to work on the brain, which accordingly remained adaptable. That is why they are at the head of the upward and onward march towards greater consciousness.

If, therefore, the mammals form the dominant branch of the Tree of Life, the primates are its leading shoot, and the anthropoids are the bud in which this shoot ends up.

Teilhard uses the simile of heat for portraying an increased degree of consciousness and comments that it is in the head of the mammals that the most powerful brains ever made by nature are to be found, and at this centre there glows, as it were, a point of incandescence.

After thousands of years rising below the horizon, a flame bursts forth at a strictly localized point.

Thought is born.

8

THE BIRTH OF THOUGHT

HAVING REACHED THE point where thought is born, Teilhard comments that man is the most mysterious and disconcerting of all the objects met with by science. In fact, he says,

We may as well admit that science has not yet found a place for him in its representations of the universe. Physics has succeeded in provisionally circumscribing the world of the atom. Biology has been able to impose some sort of order on the constructions of life. Supported both by physics and biology, anthropology in its turn does its best to explain the structure of the human body and some of its physiological mechanisms. But when all these features are put together, the portrait manifestly falls short of the reality. Man, as science is able to reconstruct him today, is an animal like the others—so little separable anatomically from the anthropoids that the modern classifications made by zoologists . . . include him with them in the same super-family, the hominidae. Yet, to judge by the biological results of his advent, is he not in reality something altogether different?

Strictly speaking the leap between animal and man biologically was extremely slight, yet it was the result of almost a revolution in the history of life—in that it contained the whole human paradox.

There is the evidence too that science, in its present-day reconstructions of the world, neglects an essential factor, or, rather, an entire dimension of the universe. To give man his true position in the world, it is necessary to consider the Within as well as the Without of things. This has already enabled us to appreciate the grandeur and the direction of the movement of life. It is our task next to divine and to describe, stage by stage, the march of mankind right down to the decisive stage reached today.

Biologists are not yet agreed on whether or not there is a design in evolution; nor is there any real agreement among psychologists, as to whether the human psychism differs specifically from that of man's predecessors or not. As a matter of fact the majority of "scientists" would tend to contest the validity of such a breach of continuity. So much has been said, and is still said, about the intelligence of animals.

Teilhard says he can see only one way to settle this question of the "superiority" of man over animals: by making straight for the remarkable phenomenon of reflection.

Reflection is, as the word indicates, the power acquired by a consciousness to turn in upon itself, to take possession of itself *as of an object* endowed with its own particular consistence and value: no longer merely to know, but to know oneself; no longer merely to know, but to know that one knows . . .

The consequences of such a transformation are immense, visible as clearly in nature as any of the facts recorded by physics or astronomy. The being who is the object of his own reflection, in consequence of that very doubling back upon himself, becomes in a flash able to raise himself into a new sphere. In reality, another world is born. Abstraction, logic, reasoned choice and invention, mathematics, art, calculation of space and time, anxieties and dreams of love—all these activities of *inner life* are nothing else than the effervescence of the newly-formed centre as it explodes on to itself . . .

If, as follows from the foregoing, it is the fact of being "reflective" which constitutes the strictly "intelligent" being, can we seriously doubt that intelligence is the evolutionary lot proper to man and to man *only*?. . . Can we . . . hesitate to admit that man's possession of it constitutes a radical advance on all forms of life that have gone before him? Admittedly the animal knows. *But it cannot know that it knows:* that is quite certain . . . It is denied access to a whole domain of reality in which we can move freely. We are separated by a chasm—or a threshold—which it cannot cross. Because we are reflective we are not only different but quite other. It is not merely a matter of change of degree, but of a change of nature.

As anticipated at the end of the last chapter, we find ourselves confronted with the genesis of the power of reflection. Outwardly, almost nothing in us had changed. But in depth, a great revolution had taken place: consciousness was now capable of perceiving itself in the concentrated simplicity of its faculties. And with man all this happened for the first time.

Those who adopt the spiritual explanation are right when they vehemently defend a degree of transcendence of man over the rest of nature. But neither are the materialists wrong when they maintain that man is just one further term in a series of animal forms.

Here, as in so many cases, the two antithetical kinds of evidence are resolved in a movement—provided that in this movement we emphasize the highly natural phenomenon of the "change of state". From the cell to the thinking animal, as from the atom to the cell, a single process (a psychical kindling or concentration) goes on without interruption and always in the same direction.

But how has the mechanism worked in its concrete reality? Had there been a witness to the crisis, what would have been externally visible to him? ... The most we have to guide us here is the resource of thinking of the awakening of intelligence in the child in the course of ontogeny ...

Life does not work by following a single thread, nor yet by fits and starts. It pushes forward its whole network at one and the same time. So is the embryo fashioned in the womb that bears it. This we have reason to know, but it is satisfying to us to recognize that man was born under the same general law. And we are happy to admit that the birth of intelligence corresponds to a turning in upon itself, not only of the nervous system, but of the whole being.

This transit to human reflection involved in the first place a change of state; then, by this very fact, the beginning of another kind of life—the interior life already referred to.

Obviously by the effect of such a transformation the entire structure of life is modified. With the advent of the power of reflection everything is changed. The phenomenon of man takes definite shape. The cell becomes someone. After the grain of matter, the grain of life, and then at last we see the grain of thought. From this crucial point life has acquired another degree, another order, of complexity. Like a river enriched by contact with an alluvial plain, the vital flux as it crosses the stages of reflection is charged with new activities.

Here again one must give the proper place to the Within of things as well as the Without. As a result of the inner energy liberated by reflection, the growth of the tree of life tends to emerge from the material organs to formulate itself in the mind. What is spontaneously mental (or spiritual) is no longer an aura round the physical; it becomes a principal part of the phenomenon. To unravel the structure of the thinking phylum, the science of anatomy by itself is not enough: it must be backed by psychology.

Teilhard describes the general direction of human growth after the transit to reflection in terms of the human element that seeks itself and grows. He says,

> We are not astonished (because it happens to *us*) to see in each person around us the spark of reflection developing year by year. We are all conscious . . . that *something* in our atmosphere is changing with the course of history.

He asks,

> How is it that we are not more sensitive to the presence of something greater than ourselves moving forward within us and in our midst?

What happens if we give the human psyche its legitimate place in the living organisms? Under the free effort of successive intelligences something irreversibly accumulates and is transmitted, at the least collectively, by means of education, down the ages. This "something" always ends up by translating itself in an augmentation of consciousness which is, in its turn, nothing less than the substance and heart of life in process of evolution.

So, from individual men there springs the human reality— the human stem. In man, considered as a zoological group, everything is extended simultaneously—sexual attraction, with the laws of reproduction; the inclination to struggle for survival, with the competition it involves; the need for nourishment, with the taste for devouring; curiosity, with its delight in investigation; the attraction of joining others to live in society.

Each of these "fibres" traverses each one of us, coming up

from far below and stretching beyond and above us. Each has contributed to the story of the whole course of evolution—of love, of war, of research, of the social sense. Each undergoes a change as it crosses the threshold of reflection—which enriches it with new possibilities, new horizons, new outlooks. Man progresses by slowly elaborating from age to age the essence and totality of a universe inherent within him; Teilhard calls this process "hominization", arising from the personalization of the individual.

He sums up this point of hominization in the following paragraph:

> The biological change of state terminating in the awakening of thought does not represent merely a critical point that the individual or even the species must pass through. Vaster than that, it affects life itself in its organic totality, and consequently it marks a transformation affecting the state of the entire planet.

Beneath the pulsations of chemistry, of the earth structure and of biology, the same fundamental process is detected—one which was given material form in the first cells and was continued in the construction of the nervous systems. Reflection and the birth of the "psychological" follow.

At this point Teilhard envisages another and higher function of the mind which he describes as *noogenesis* leading to the *noosphere*. This has been translated as the gradual evolution of mind or mental properties, while Teilhard often uses it in the sense of the "field of human thought" or "climate of mankind's opinion".

> Much more coherent and just as extensive as any preceding layer, it is really a new layer, the "thinking layer", which, since its germination at the end of the Tertiary period, has spread over and above the world of plants and animals. In other

48

words, outside and higher than the biosphere there is the no-osphere.

He considers the development of human reflection on this scale—perceiving itself in its own mirror, as it were, a decisive factor for our understanding of the earth.

The noosphere, Teilhard claims, outstrips in evolutionary importance the already mentioned geological spheres, i.e. barysphere (metallic); lithosphere (rock); hydrosphere (water); biosphere (living); to say nothing of the atmosphere and stratosphere.

With hominization and the noosphere the earth gets a new sphere or better still it finds its soul.

Among all the stages successively crossed by evolution, the birth of thought comes directly after, and is the only thing comparable in order of importance to, the condensation of the terrestrial chemism or the advent of life itself . . .

This sudden deluge of cerebralization, this biological invasion of a new animal type which gradually eliminates or subjects all forms of life that are not human, this irresistible tide of fields and factories, this immense and growing edifice of matter and ideas—all these signs that we look at, for days on end—exist to proclaim that there has been a change on the earth and a change of planetary magnitude.

To a Martian visiting this planet today, if he were capable of analysing the psychic as well as the physical manifestations of the earth, the first characteristic that would be noticeable would be this "phosphorescence" of thought.

The greatest revelation open to science today is to perceive that everything precious, active and progressive originally contained in that cosmic fragment from which our world emerged, is now concentrated in a "crowning" noosphere.

9

MAN'S ORIGINAL FORMS

THE MORE WE find of fossil human remains and the better we understand their anatomic features and their succession in geological time, the more evident it becomes that the human species (however important the position reflection gives it), did not, at the moment of its advent, make any sweeping change in nature.

In fact, it emerged like any other species. It took its shape gropingly as it were, in the midst of similar forms. As Teilhard puts it, "Man came silently into the world".

We first catch sight of him when, revealed by indestructible stone instruments, we find him sprawling all over the old world from the Cape of Good Hope to Peking. At that point, without doubt, he already spoke (or at least grunted) and lived in groups and made fire. In terms of science the first man is, and can only be regarded as, *a crowd*, and his infancy is made up of thousands and thousands of years.

This fact is inevitably disappointing—leaving our curiosity unsatisfied; for naturally we would be interested in knowing what happened during those first thousands of years. Still more, what marked the first critical moment—especially, we would love to know what those first parents of ours looked like—the ones that stood just this side of the threshold of reflection.

That threshold, as Teilhard has already said, had to be

crossed in a single stride. But no photographic record of such a passage to this stage of reflection could possibly be available, for the simple reason that the phenomenon took place inside the original shoot or peduncle. Although the tangible form of this transition escapes us, we can, at least indirectly, form an opinion as to its complexity and initial structure.

A number of anthropologists believe that the peduncle of our race must have been composed of several distinct but related bundles. According to them man must have started simultaneously in several regions during the Pliocene era, some ten million years ago, during which there were known to be several manlike forms which represented offshoots from the main human tree. Teilhard refers to these developments as a whole series of points of hominization scattered along a sub-tropical zone of the earth and hence several human stems becoming genetically merged somewhere below the threshold of reflection; not a focus but a "front" of evolution.

At the same time, Teilhard himself is inclined towards a slightly different hypothesis. He feels inclined to minimize the effects of parallelism in the initial formation of the human branch. He prefers to align himself, if not with the proposition of a single initial couple (i.e. monogenism, which the science of man can say nothing directly for or against), but rather with that of a single phylum or stem, a conception which he has fully developed in the earlier part of his book. He adds, moreover, that if this is right, all human lines join up genetically—but fundamentally, at the very point of reflection.

On this assumption, Teilhard then examines the probable formation of such a stem in greater detail. When, in fact, did this phenomenon take place in the human time table?

And, what would be the minimum number of individuals who must have undergone the genesis of reflection? Could such a change be compared to a diffuse current in a river—or propagate itself like crystallization beginning with a few particles?

In answer to these questions he makes the two following points.

First, that on every hypothesis, however solitary his advent, man emerged from a general groping of the world. He was born a direct lineal descendant from a total effort of life, so that the species has an axial value and a pre-eminent dignity. At bottom, to satisfy our intelligence and the requirements of our conduct, we have no need to know more than this.

The second point is that, fascinating as the problem of our origin is, its solution even in detail would not solve the problem of man. We have every reason to regard the discovery of fossil men as one of the most illuminating and critical lines of modern research. We must not, however, on that account, entertain any illusions concerning the limits in all its domains of the form of analysis that we call embryogenesis. If in its structure the embryo of each thing is fragile, fleeting, and hence, in the past, practically ungraspable, how much more is it ambiguous and indecipherable in its lineaments? It is not in their germinal state that beings manifest themselves, but in their flowering. Taken at the source, the greatest rivers are no more than narrow streams.

To grasp the truly cosmic scale of the phenomenon of man, we had to follow its roots through life, back to when the earth first folded in on itself. But if we want to understand the specific nature of man and divine his secret, we have no other method than to observe whatever reflection has already provided and indicates may be ahead.

IO

MANKIND IN THE MAKING

MAN'S PREDECESSORS ARE to be found in the Neanderthal and Peking man. Teilhard himself was in China on the
expedition which discovered evidence of the latter, known
scientifically as Sinanthropus. Prior to this, came Pithecanthropus of Java, originally represented only by the simple
skull, but later by more satisfactory specimens. Teilhard
points out that both these types were definitely hominid in
their anatomy, i.e. near man.

Compared to the great apes there is a considerable biological difference between them and man, but they fall
naturally into the same cast, with a relatively short face and
a relatively short head. At any rate, there are quite definite
signs that these are on the human side of the line between
apes and man.

All things considered, the scientist can affirm without hesi-
tation that a further evolutionary stage towards man has
been reached through these hominid types which represent
the period through which the modern man must have once
passed in the course of his evolution.

Although these types were found in the farthest confines
of Eastern Asia taken as a whole, they obviously belonged
to a very much bigger anthropological species. Through
them, in fact, we are able to envisage a whole wave of mankind. So far, at any rate, we can find no trace of industry

associated with them, but this is due to the conditions in which they were found. However, Sinanthropus was caught in his lair, a filled-up cave littered with stone implements mixed with charred bones, and it is known that Sinanthropus worked with stone tools and lit fires.

Teilhard points out that these two activities do point to the beginning of the level of reflection. If this is really so, these pre-hominids were psychologically much nearer to us than may have been supposed. It must have taken time to discover fire and the art of making a cutting tool. Teilhard states that during the period there were five different types of hominid which were certainly related to each other and in his view these developments in China and Malaya might well have been paralleled elsewhere, further west.

He considers however that at the same time in more central regions of the world the elements of a new and more compact wave of mankind were mustering ready to take over from the primitive types.

Neanderthal Man

When the curtain rises again some sixty thousand years ago, and we can see the scene once more, we find that the pre-hominids have disappeared. Their place is now occupied by the Neanderthaloids.

This new human type is much better represented by fossil remains than its predecessors, being both more recent, and more numerous. Little by little the network of thought has extended and consolidated.

We find both progress in number and progress in hominization.

By now we have perhaps true man, but man who was not yet obviously us. At this stage there were two distinct groups of the Neanderthal man at different levels of evolution. One was in the process of dying out and the other, the pseudo-Neanderthal, in the process of evolving.

Teilhard compares the pre-hominids and the two sorts of Neanderthal man to the pattern of a tree where some leaves have just fallen, some leaves are still alive but beginning to turn yellow, and others have not yet opened but are full of vigour.

Homo Sapiens

Now sweeping over the Neanderthaloids comes the sudden invasion of Homo Sapiens driven by climate or the restlessness of his soul. Where did he come from, this new man? Some anthropologists suggested that he is a direct descendant from Sinanthropus, but Teilhard views things in another way. He maintains that somewhere or other, and in his own way, Upper Palaeolithic man must have passed through a pre-hominid phase and then through a Neanderthaloid one, but like the mammals he disappears from our field of vision in the course of his evolution.

It is Teilhard's view that the law of succession was operating through this channel. You could say a newcomer emerged triumphantly one fine day in the midst of those pseudo-Neanderthaloids, whose vital and ancient stem we have already mentioned.

Now the man we find on the face of the earth at the end of the Quaternary period is already modern man. The palaeontologist no longer finds it easy to distinguish between the remains of these fossil men and men today. Thirty thousand years is a long period measured in the terms of our lifetime, but it is a mere second in the terms of evolution.

During that period there is no appreciable breach of continuity in the human phylum; no major change in the progress of its physical properties. This is where we get our greatest surprise. It is only natural that the stem of *Homo sapiens fossilis* studied at its point of emergence, far from being simple, should display in the composition and divergence of its fibres the complex structure of a fan. This is, we know, the initial condition of every phylum on the tree of life. At the least, we should have expected a cluster of relatively primitive forms, something antecedent to our present races. But we find the opposite. At the point of emergence we find evidence of types identical to those living today in approximately the same regions of the earth: negroes, white men, yellow men and those various groups already settled north, south, east and west in their present geographical zones. These are the types we would have found all over the ancient world from Europe to China at the end of the last Ice Age.

So, when we study the Upper Palaeolithic man it is really ourselves and our own infancy we are finding. Not only is the skeleton of modern man already there, but the frame of modern humanity. We see the same general bodily form, the same fundamental distribution of races, the same essential aspirations in the depths of their souls.

Most of Upper Palaeolithic man's intelligence however seems to have been used up in the sheer effort to survive and reproduce. We have no idea of what went on in the minds of those distant cousins of ours. Where we could not be mistaken is in perceiving in the artists of those distant ages the power of observation, a love of fantasy and a joy in creation. A flower of consciousness not merely reflecting upon itself, but rejoicing in doing so.

Although these were not in the fullest sense adult present-day men, we are looking at men who have reached the age of reason. Their brains were already perfect, and since that time there seems to have been no measurable variation or increased perfection in man's instrument of thought.

Does this mean that evolution in man ceased with the end of the Quaternary era? Not at all. But since that date evolution has merely overflowed its anatomical characteristics to spread or perhaps even to transplant its main thrust into the zones of consciousness both individual and collective. It is the course of evolution in that form that we shall almost exclusively be following.

II

THE NEOLITHIC AGE

SOCIAL DEVELOPMENT OF the higher animals is a progression that comes relatively late. In man, however, for reasons connected with his power of reflection, this transformation has been accelerated. As far back as we have gone, our great-great-ancestors are found in groups and gathered round a fire. But in the Neolithic Age (late Stone Age) the great cementing of human elements began. It has never stopped since.

This age was a critical one and ranks high among all the periods of the past, for in it civilization was born.

There is a gap in archaeological discovery between the levels of chipped stone and the first levels of polished stone and pottery, a gap which is negligible in terms of the vast eras of geology, but still long enough for the selection and domestication of all the animals and plants on which we are still living today. On this side of the gap we find sedentary and socially organized men in the place of the nomadic hunters of the horse and the reindeer. In a matter of ten or twenty thousand years man divided up the earth and struck his roots in it.

In this decisive period of socialization, as previously at the instant of reflection, a series of partially independent factors seems to have converged mysteriously to favour and even to force the pace of hominization.

First come the incessant advances of multiplication. With

the rapidly growing number of individuals, the available land diminished. Groups pressed against one another. Under this pressure the idea was born of conserving and producing on the spot food and other necessities previously sought for far and wide. Agriculture and stock breeding, the husband-man and the herdsman, replaced mere gathering and hunting.

From that fundamental change all the rest followed, even communal and juridical structures whose vestiges we can still see today in the great civilizations of the world. As regards property, morals and marriage, every conceivable social form seems to have been tried.

It was a marvellous period of investigation and invention when, in the unequalled freshness of a new beginning, the eternal groping of life burst out in conscious reflection. Practically everything seems to have been attempted: the selection and improvement of fruits, cereals and livestock; the science of pottery; weaving; and very soon, the first elements of symbolic writing and the beginnings of metallurgy.

Man was in his full flush of expansion. It was at the turn of this age that man reached America, passing through Alaska, then free of ice. Many other routes were also explored, followed by the discovery of new materials. Among these men were many hunters and fishermen still living a more or less Stone Age life despite their pottery and polished stone. But besides these there were genuine tillers of the soil including the maize eaters. At the same time, no doubt, another human layer began to spread, whose long trail is still marked by the presence of banana trees, mango trees and coconut palms. This was the fabulous adventure westward across the Pacific.

At the end of this transformation, whose existence we can

only infer from the results, the world was practically covered with a population whose remains—polished stone implements, millstones, etc.—litter the old earth of the continents. Mankind was of course still very much split up. To get an idea of this we have to appreciate that the first white men in America or Africa found a veritable mosaic of groups profoundly different both ethnically and socially.

But mankind was to some extent already linked up. Since the age of the reindeer the peoples had been little by little finding their definite place in the world. Between them commerce increased as well as the transmission of ideas. Traditions became organized and a collective memory was developed. This was the beginning of the Noosphere, the sphere of communal consciousness, in its encirclement of the earth.

The Rise of the West

Teilhard points out that there is, of course, a distinction between history, supported by written and dated documents and the pre-history on which we have to base most of our findings, but there is no real breach of continuity between the two. History records the normal expansion of Homo Sapiens following and created by the Neolithic evolutionary period. There are various stages of progress described by Teilhard in the development of different stems: some disappearing, causing a thinning out, some fresh buddings which make the foliage more dense. Some branches wither, some sleep, some shoot up and spread everywhere. He regards the whole series of situations and appearances as similar to those met with in any phylum in a state of active proliferation.

In Teilhard's view, so long as science had to deal only with pre-historic, more or less isolated, human groups, the general rules of animal evolution were still approximately valid.

But from that point on there is more emphasis needed on the complex scale of man's varying groupings, his geographical distribution, economic links, religious beliefs and social institutions.

There is no need to emphasize the reality, diversity and continual germination of human collective unities. Potentially at any rate these of course incorporate the roots of divergence. Birth, the multiplication and evolution of nations, states and civilizations all illustrate the unifying/ divergent thrust.

But there is one thing that must not be forgotten if we want to enter into and appreciate the drama. However hominized the events, the history of mankind in this rationalized form really does prolong—though in its own way and degree—the organic movements of life. It is still natural history through the phenomena of social ramification that it relates.

In the case of the animal branches or phyla with low endowment of consciousness, reactions are limited to competition and eventually to elimination. The stronger supplants the weaker and ends by stifling it.

With man (at all events with Post-Neolithic man) simple elimination tends to become exceptional. However brutal the conquest, suppression is always accompanied by some degree of assimilation. Even when partially absorbed, the vanquished still reacts on the victor so as to transform him.

Under this twofold influence true biological combinations are established and fixed. Formerly, on the tree of life we had a mere tangle of stems; now over the whole domain of homo sapiens we have synthesis.

Because of the haphazard configuration of the continents of the earth, some regions are more favourable than others

for the concourse and mixing of races; extended archipe-
lagoes, junctions of valleys, vast cultivable plains, particularly those irrigated by a great river. In such places there has
been a natural tendency for the human mass to concentrate.

Five of these can easily be picked out—Central America,
the South Seas, the basin of Yellow River, the valleys of the
Ganges and the Indus, and the Nile Valley and Mesopotamia.
The people of the last three probably appeared at the same
period, but the first two were much later. They were all
largely independent of one another, each struggling to
spread and ramify. History seems to consist in the conflict
and eventual harmonization of these great currents.

In the main, the contest for the future of the world was
fought out by the agricultural plain dwellers of Asia and
North Africa. One or two thousand years before our era the
odds between them may have seemed fairly equal. But today we can see that there were the seeds of weakness in the
East. The old China lacked both the inclination and the
impetus for deep renovation, and scarcely changed in ten
thousand years. The population was fundamentally agricultural with the emperor the biggest owner. It was a
population specializing in brick work, pottery and bronze,
the study of symbolic signs, and the science of the stars. This
was a refined civilization, certainly, but one with unchang-
ing methods and little consequent capacity for regeneration.

While China concerned herself with building up a science
of physics, India allowed itself to be drawn into a philosophic and religious maze only to become lost there. Her
people were incapable of building the world.

Step by step we are driven nearer to the more western
zones, to the Euphrates, the Nile and the Mediterranean,
where reason was harnessed to facts and religion in action.

Mesopotamia, Egypt, Greece, with Rome soon to be

added—and above all the mysterious Judaic–Christian ferment which gave Europe its spiritual form—these are the areas of movement we are led to. An ever more highly organized consciousness of the universe passed from man to man and became steadily brighter.

THE MODERN EARTH

IN EVERY EPOCH man has thought himself at a turning point of history and, says Teilhard, to a certain extent the belief was true. For man is ever advancing on a rising spiral. Nevertheless there are times when this tendency is accentuated, and Teilhard believes that a change of profound importance is in fact taking place at this very time.

During the last two centuries, at any rate, we have entered a basically different world. Before then our civilization was largely based on the partition of the land, the arable soil. Bit by bit this process has changed, and today the wealth of nations has little in common with their frontiers. The machine age has changed the face of industry, and consequent social changes have resulted in the awakening of the masses.

What is now troubling us intellectually, politically, and even spiritually, is something quite simple. We are passing through a change of age. As yet the new age has no new name; the future will decide whether we call it the age of industry or the age of oil, electricity or the atom; or, perhaps, even, the age of science.

The words matter little. What does matter is that life is taking a decisive step in a new direction—a critical change in the communal consciousness, or in the noosphere, of which Teilhard has already written. Our world of factories and offices seething with work and business, this great organism lives in the final analysis for the sake of a new soul.

Beneath a change of age lies a change of thought. Where are we to look for it, where are we to locate this subtle alteration which, without appreciably changing our bodies, has made new creatures of us? In one single direction, in a new intuition involving a total change in the face of the universe; in other words, in an awakening, we have become conscious of the movement which is carrying us along, and have therefore realized the formidable problems set us by this reflective expression of human efforts.

In the first stages in man's awakening to the immensities of the cosmos, space and time still remained independent of each other; they were two great elements quite separate one from the other, extending infinitely, no doubt, but in which things floated about or were packed together in ways owing nothing to the nature of their setting.

It was only in the middle of the nineteenth century that the light dawned at last, revealing the irreversible coherence of all that exists. It was only then that it was realized that time and space are organically joined to make the stuff of the universe.

That is the point we have now reached, and that is how we perceive things today. In the last century and a half, the most prodigious event perhaps ever recorded by history, since the threshold of reflection, has been taking place in our minds: the definitive opening up of consciousness to a scale of new dimensions; and in consequence the vision of an entirely different universe, without any change of line or features—by this simple transformation of our outlook on time and space.

Now we are becoming capable of seeing it, in terms not of time and space alone, but also of duration—or, more specifically, "biological space-time".

Having commented on the effect of space-time and duration in the universe, Teilhard goes on to define what in his opinion is the modern man. He says that it is surprising how naïvely naturalists and scientists continued, in their discoveries, to stand outside the universal stream of which they were obviously part.

Subject and object were separated from each other in their eyes. It did not occur to the first evolutionists that their own scientific intelligence had anything to do in itself with evolution. But how, he asks, could we incorporate thought into the flux of space-time without being forced to grant thought itself first place in the process of evolution itself. He borrows here Julian Huxley's "striking expression"—man discovers that he is nothing else than *evolution become conscious of itself.* This view in Teilhard's opinion is destined to become as instinctive and familiar to us as that of a third dimension in space. It is a new light.

Having followed step by step, from the early earth onwards, the successive advances of consciousness we can now look, backwards as it were, at the pattern as it has so far developed. From this viewpoint, Teilhard claims, a triple unity persists and develops—the unity of structure, the unity of mechanism and the unity of movement.

Unity of Structure

Fanning out: this is the pattern we see on the tree of life. We found it when considering the origins of mankind and of the principal waves of humanity. We see it today in the complex ramifications of nations and races.

And now we shall be able to discern the same pattern again in forms which are closer to us, though more immaterial. In space-time, is there such a great difference between an animal spreading its limbs or equipping them with feathers, and a pilot soaring on wings which man has had the ingenuity to equip him with? Evolution is linked with a thousand social phenomena which we should never have imagined were linked with biology. It would not directly be possible, for example, to identify the development of new industries, or of philosophic and religious doctrines as a part of the process of evolution.

Unity of Mechanism

Teilhard refers to the mutations which occurred when he described the appearance of successive zoological groups. We have mutations narrowly limited round a single focus and mutations in which whole blocks of mankind are involved. In some of these cases we see the process in full view and can detect the progressive leaps of life in an active and complete way.

Invention also can legitimately be regarded as an extension of the mechanism whereby each new form has always germinated on the tree of life. In the same sense the instinctive gropings of the first cells link up with the learned gropings of our laboratories. Both exhibit the same mechanism.

The process of "trying all and discovering all" comes to us from "far away" and started at the same time as the light from the first stars. "The spirit of research and conquest is the permanent soul of evolution."

Unity of Movement

Man is not the centre of the universe, as once we thought in our simplicity, but something much more wonderful. He is the arrow pointing the way to the final unification of the world in terms of life. Man is the last-born, and the most complicated, of all the successive layers of life.

It is true that we do not yet know how characters are

formed in the secret recesses of the germ cells. Neither has biology yet found a way of reconciling the spontaneous activity of individuals with the blind determination of the genes and chromosomes. But, then, Teilhard asks, what becomes of the part played by the power of invention?

In our innermost being, he says, we feel the weight of a definite legacy of life handed down to us from the past. Depending on how industriously we use our powers of vision, we see too the further advance of this vital wave. This is the collective memory and intelligence of human life.

The further the living being emerges from the anonymous masses by the radiation of his own consciousness, the greater becomes that part of his activity which can be transmitted by means of education and imitation. This layer of heredity becomes a self-reflecting organism in which the evolution of the tribe merges with that of the individual.

The Problem of Action

In commenting further on the change of outlook due in part to the recognition of biological space-time, Teilhard uses the analogy of a child who is terrified upon opening his eyes for the first time. It is as though one had emerged from a dark prison and found oneself dazed and giddy at the top of a tower. In his opinion the extent of modern disquiet is linked with our confrontation of space-time.

Something seems to threaten us without our being able to say what. The novelty of space-time, in Teilhard's view, is responsible for this feeling of futility. The enormity of space presents us with a frightening prospect. Does not the very thought of galaxies whose distance apart runs into hundreds of thousands of light years cast a gigantic shadow over our beliefs of the past?

Time and space are indeed terrifying when frankly regarded. But there is the other side of the picture—namely, that we can perceive an evolution which animates these enormous dimensions. Time and space themselves become humanized as soon as a comprehensible direction appears.

Half our present uneasiness would be turned into happiness if we could once make up our minds to accept the facts and place our modern universe within a communal layer of thought. The solution lies in our new knowledge of evolution, and in our belief that there is an outcome to that evolution.

When the first spark of thought appeared upon the earth, life found that it had brought into the world a power capable of criticism and judging the conditions in which it lived. We began to become aware that in this great game of life we were the players as well as being the cards and the stakes.

The last century has witnessed a realization of this new ability to rid ourselves of our previous servitude. In the future there will surely be further crises in the development of the communal consciousness of the world.

Teilhard says that we must be assured of the opportunities and the chances to fulfil ourselves and thus be enabled to progress to the utmost limits of ourselves. In Teilhard's opinion, consciousness is a dimension which has no limits.

There are innumerable critical points on the way, but a halt or a reversion is impossible, for the simple reason that every increase of internal vision is essentially the germ of a further vision which includes all the others and moves still further on.

The more man becomes man, the less will he be prepared to move except towards new horizons. If that were not so, all movement of the communal consciousness on earth would virtually be brought to a stop.

Critical minds make the mistake of believing that life would continue its peaceful cycle when deprived of light, of hope and the attraction of an inexhaustible future. Without the taste of life, mankind would soon stop inventing and constructing for work it knew to be doomed in advance.

Once we are reduced to saying, "What's the good of it all?" our efforts would flag, and with that the whole of evolution would come to a halt; because *we are evolution*.

Teilhard says,

Life, by its very structure, having once been lifted to its stage of thought, cannot go on at all without requiring to ascend ever higher.

This is enough for us to be assured of the two points of which our action has immediate need.

First, that there is for us, in the future, in one form or another, not only survival but also *super-life*.

Second, that to imagine, discover and reach this super-life, we have only to think and to walk further in the direction in which the evolutionary process takes us.

13

THE COLLECTIVE ISSUE

IT IS RARE indeed for extreme individualism to go beyond
the bounds of a philosophy of its own immediate enjoyment
and feel the need to come to terms with the greater requirements of communal action.

"Progress by isolation" has in our times fascinated large
sections of mankind—the doctrine of racial selection and
election is an example. Collective egotism is more easily
aroused than individual egotism. The history of the animate
world shows us a succession of groups springing up one after
another, one on top of the other, through the success and
domination of a privileged group. Why should we be
exempt from this general rule? Why should there not be a
struggle for life and the survival of the fittest, the trial of
strength? Like any other stem, will not super-man be an off-
shoot from the single bud of mankind?

Both isolation of the individual and of the group seem to
produce a plausible justification for life; but we shall see,
however intense the passions they stir, that these are cynical
and brutal theories. Sometimes we may not be able to help
being responsive to some of the calls to violence fundamentally isolationist causes involve, but they are nevertheless
a sinister departure from a great truth.

It is necessary to see clearly that individual and group

isolationists deceive themselves, and render it biologically impossible to form a true spirit of the earth.

The Confluence of Thought

Throughout *The Phenomenon of Man* Teilhard has emphasized the fact that the elements of the world are able to influence and mutually penetrate each other by reason of their Within. This mutual penetration grows and becomes increasingly perceptible in the case of organized beings, and finally in man it reaches a maximum degree of consciousness. It is not by accident that the earth is round; because of its topographical limitations it is like a gigantic molecule.

If humanity had been free to spread indefinitely on an unlimited surface, something altogether different from the modern world would have resulted. Originally there was no serious obstacle to the human waves expanding over the surface of the globe, but in due time these waves began to recoil upon themselves, all available amenities eventually being occupied and populations having to pack in tighter and tighter. In this way we have come to our present position as an almost solid mass of "hominization".

The human elements infiltrate more and more into each other, their minds mutually stimulated by proximity. Owing to the railway, the motor car and the aeroplane, the physical influence of each man, formerly restricted to a few miles, now extends to hundreds or more. And with the advent of radio, each individual finds himself figuratively present in every corner of the earth.

Mankind finds itself submitted to ever more intense pressure, and each advance causes a corresponding expansion throughout the steadily increasing population of the world. This coalescence, or coming together, of the forces of the world results in a corresponding concentration of the energies of consciousness. Teilhard has already drawn attention to this increasing mixing together of mankind in political social groups and has hinted at the ultimate significance of this phenomenon.

There comes a point when each divergent branch of the human race is no longer able to separate. The initial phylum as it spreads out, remains intact, like a gigantic leaf, fanning out in a characteristic expansion. In man we find on every level a blending of races in civilization or political bodies. Zoologically speaking, mankind offers the unique spectacle of a species capable of an achievement which no previous species has come near to: the coalescence or uniting together of the entire phylum.

Teilhard compares this coalescence of living branches with the serried shoots of ivy. Until man came, the most life had managed to realize in the matter of association had been the gathering socially together of the finite extremities of the same phylum.

This resulted in essentially family groups, created in a purely functional need for house-making, defence or propagation—such as the beehive or the ant heap—limited, in fact, to the offspring of a single mother.

But, from man onwards, thanks to the universal framework of thought, free rein is given to the forces of confluence. At the heart of this new milieu, the branches themselves become welded together even before they have managed to separate off.

Just as when the meridians of a sphere separate at one pole and come together at the other, so divergence gives way to a movement of convergence in which races, peoples and nations consolidate, unifying one another by mutual fecundation.

This process of "human planetization", of cohesion, this

furling back upon itself of a "bundle" of potential species around the surface of the earth, becomes the main feature of mankind's evolution.

Why should there be unification and what purpose does it serve?

Teilhard answers this question by setting side by side two equations which, he says, have gradually been formulating themselves from the moment we began trying to place the phenomenon of man in the world.

$$
\begin{aligned}
\text{Evolution} &= \text{Rise of Consciousness} \\
\text{Rise of Consciousness} &= \text{Union Achieved}
\end{aligned}
$$

These equations become more intelligible as soon as we see the conditions they define as the natural culmination of an unvarying cosmic process activated continually by the inter-related actions of the Within and the Without of the earth.

First, the molecules of carbon compounds with their thousands of atoms symmetrically grouped; next the cell, which, within a very small volume, contains thousands of molecules linked in a complicated system. Then the metazoa, in which the cell is no more than an almost infinitesimal element, and later, the manifold attempts made sporadically by the metazoa to enter into symbiosis (mutual association) and raise themselves to a higher biological condition. And, finally, the "thinking layer" which develops and intertwines its fibres to create the living unity of a single tissue.

Teilhard describes this gigantic operation as a sort of megasynthesis (or super arrangement) to which all the thinking elements of the earth today find themselves individually and collectively subject, and as a combination of tangential and radial energies along the principal axis of evolution—ever more complex, ever more conscious.

In such a situation the vital error of an egocentric ideal of the future is very evident. The idea of "everyone for himself" is both false and against nature.

The outcome of the world, the gates of the future, the entry into the superhuman—these are not thrown open to a few privileged, nor to one chosen people, to the exclusion of all others.

They will open only to an advance of all together.

The Spirit of the Earth

Mankind was first a vague entity, felt rather than thought out, by which an obscure feeling of perpetual growth was allied to a need for universal fraternity. No one can escape being haunted or even dominated by the idea of mankind.

In the course of a few generations all sorts of economic and cultural links have been forged around us and they are multiplying in geometrical progression. Nowadays, over and above the bread, which to simple Neolithic man symbolized food, each man demands his daily ration of iron, copper and cotton, of electricity and oil, of discoveries, and of international news. The whole earth is required to nourish each one of us.

Is this not like some great body being born—with its limbs, its nervous system, its perceptive organs, its memory —the body, in fact, of that great thing which had to come to fulfil the ambitions aroused in the reflective being by the newly acquired consciousness?

No evolutionary future awaits man except in association with all other men. The dreamers of yesterday glimpsed this, and, in a sense, we see the same thing. But what we are better able to perceive are the cosmic roots, the particular physical substance, and the specific nature of mankind of

75

which they would have had only a presentiment—but which, unless we shut our eyes, we cannot overlook.

For us, more aware of the dimensions and structural demands of the world, the forces which converge upon us from without, or arise from within and drive us ever closer together, are losing any semblance of chance happenings.

Mankind becomes a consistent whole as soon as it is brought within the compass of biological space-time and appears amongst other equally vast realities as a continuation of the lines of the universe.

Admittedly it is difficult for man to visualize this whole correctly—for it is necessary to envisage, besides the individual realities, the collective ones. For that reason, Teilhard says he has represented the movements of life as concepts, such as phyla, layers, branches, etc.

Once the eye has become adjusted to the perspectives of evolution, these specific groups become as clear as an ordinary isolated object. And it is in this particular class of dimensions that mankind takes its place.

Being a collective reality, mankind can be understood only in so far as we are able to recognize the type of conscious synthesis which emerges from the process of its unification. In the last resort it is definable only as a mind.

Teilhard then divides the manner in which we can picture the form mankind will assume in the future, into either the terms of science, or those of superconsciousness.

First, science. The future of science is outlined on our horizon, as a first approximation, as the establishment of an over-all and completely coherent perspective of the universe. Intellectual discovery and synthesis will be no longer mere speculation but creation. Those who envisage the crown of evolution as a supreme act of collective vision are largely right. The march of humanity develops inevitably in the direction of the conquest of matter in the service of the mind —resulting in increased power for increased action, and finally and above all increased action for increased being.

There is more to existence nowadays than the search for gold. It is the search for life that should occupy much of mankind. Soon we expect to be able to control the mechanism of organic heredity. Our own thought may eventually be able to perfect artificially the thinking instrument itself.

The dream upon which human research feeds is that of mastering the ultimate energy of which all other energies are merely servants; and thus, by grasping the very mainspring of evolution, seizing the tiller of the world.

However far science pushes its discovery of what Teilhard describes as "the essential fire", it will always find itself in the end facing the same problem—namely how to give to each and every element its final value by classifying them within an organized whole.

This brings him to the second way of looking at the future: in terms of the superconscious. He reverts to his conception of mega-synthesis, a word which should be understood as applying to the sum of all human beings. Claiming that the universe is necessarily homogeneous in its nature and dimensions, he says that the still unnamed thing which the gradual combination of individuals, peoples and races will bring into existence, must needs be super-physical if it is to be coherent with the rest.

He goes on to say that the stuff of the universe has not yet completed its evolutionary cycle, and that we are therefore moving forward towards some critical point that lies ahead. The noosphere tends to constitute a single enclosed system in which each element sees, feels, desires and suffers for itself the same things as all the others at the same time.

We are faced with a harmonized collectivity of consciousness equivalent to a sort of super-consciousness. The idea is that of the earth becoming covered not only by myriads of grains of thought but by a single thinking envelope. This would form, functionally, no more than a single vast grain of thought on a sidereal scale, the total of reflecting individuals grouping themselves together and reinforcing one another in the act of a single unanimous reflection.

This is the general form in which, by analogy and in symmetry with the past, we are led scientifically to envisage the future of mankind. To the common sense "man in the street", and even to that philosophy of the world in which nothing is possible save what has always been, perspectives such as these will seem highly improbable. But to a mind become familiar with the fantastic dimensions of the universe they will, on the contrary, seem quite natural, because they are simply proportionate with the astronomical immensities.

In the light of this, it is possible to find a relatively simple meaning for the troubles which now disturb the noospheric layer of mankind. Due to the combined influence of mechanization and the enhancing of communal thought, we are witnessing a formidable upsurge of unused powers.

Modern man no longer knows what to do with the time and potentialities he has unleashed. We groan under the burden of wealth, we are haunted by the fear of unemployment. When we consider the increasing compression of the elements at the heart of free energy, we can scarcely fail to see in this twofold phenomenon symptoms of a leap forward of radial energy—a new step in the genesis of mind.

As things are now going, it will not be long before we run full tilt into one another. Something will explode if we persist in trying to squeeze into our inadequate accommodation the material and spiritual forces that are making themselves felt in the world today.

A new domain of expanding consciousness—that is what we lack. It is staring us in the face, if we would only raise our heads to look at it.

Peace through conquest, work in joy. These are waiting for us beyond the line where empires are set up against other empires ... in the unanimous construction of a *spirit of the earth*. How is it then that our first efforts towards this great goal seem merely to take us farther from it?

14

THE HYPER-PERSONAL

NINETEENTH-CENTURY MAN lived in sight of a promised land. It was thought then that we were on the threshold of a golden age under the umbrella of the new sciences. Instead, we found ourselves slipping back into a world of ever more tragic dissension.

With the evolution of man stretching over half a million years, it is perhaps not so very surprising to find man still at loggerheads with himself. We have to keep things in proper perspective. Planetary movement involves planetary majesty: we cannot expect the earth to transform itself in the space of a generation or two. Man's advance must inevitably be almost imperceptible. Sometimes it may seem that we are up against a dead end.

There are, no doubt, formidable pressures which hem in human individuals and peoples, forcing them up against one another geographically and psychologically. There are imponderable currents which make us all the slaves of the obscure seethings of the human mass. Of course, we must react to such conditions; but with the satisfaction of knowing that they are not only the sign of progress, but the price paid for it.

Teilhard says that at no period in history has mankind been so well equipped to reduce its multitudes to order. Yet, instead of the upsurge of consciousness which one might have expected, it is mechanization that seems to emerge. *Eppur si muove!*—and yet it does move.

But, he says, like the engineer when a source of energy runs amok, far from questioning the power itself, we should simply turn to and work out our calculations afresh. An up-surge of consciousness can be achieved, if we give due place to the person and the forces of personalization.

The Convergence of the Person

Modern man, according to Teilhard, is far too obsessed with the need to depersonalize all that he most admires, for two reasons.

The first is his insistence on analysis, which, marvellous though it is as an instrument of scientific research, leaves us, after breaking down synthesis after synthesis, confronted with a pile of dismantled machinery. The other reason lies in our discovery of the sidereal world, so vast that it seems to do away with all sense of proportion between our own being and the dimensions of the cosmos around us. Only one reality seems to survive capable of spanning both the infinitesimal and the immense—namely energy, the new spirit, the new god.

Under the influence of such impressions as these, it looks as if we have lost both respect for the person and the under-standing of his true nature. Yet, if we try to pursue the logic and coherence of facts to the very end, we seem to be led to an opposite view by the notions of space-time and evolution.

That evolution is an ascent towards consciousness is no longer greatly contested. If it is so, it should culminate in some sort of supreme consciousness.

There is a threefold property possessed by every consciousness: (1) of centring everything partially upon itself; (2) of being able to centre itself upon itself constantly; and

(3) of being brought more into association with all the other centres surrounding it.

As regards the apparent opposition between the All and the Person our difficulties would be resolved if only we understood that, by structure, the noosphere represents a whole that is not only enclosed, but also centred. Because space-time contains and engenders consciousness it is necessarily of a convergent nature.

The more immense the sphere of the world, the richer and deeper, and hence the more conscious, is the point at which the "volume of being" that it embraces is concentrated; because the mind, seen from our side, is essentially the power of synthesis and organization.

Far from being mutually exclusive, the Universal and Personal grow in the same direction and culminate simultaneously in each other.

The Personalizing Universe

What is the work of works for man, if not to establish in each one of us an absolutely original centre in which the universe reflects itself in a unique way?

And those centres are our very selves, the very centre of our consciousness. This "essence" is obviously not something of which we can dispossess ourselves for the benefit of others, as we might give away a coat or pass on a torch. For we are the very flame of that torch. To communicate itself, my ego must abandon itself or the gift will fade away. The conclusion is inevitable that the concentration of a conscious universe would be unthinkable if it did not reassemble in itself *all consciousness* as well as all the conscious—each particular consciousness becoming still more itself.

This is equally true in any domain—whether it be the cells of a body or the members of a society. The parts of every organized whole perfect and fulfil themselves in that whole.

Ultimately, there can only be a distinct centre radiating at the core of a system of centres; a grouping in which personalization of the All and personalization of the elements reach the maximum simultaneously and without merging under the influence of a supremely autonomous focus of union.

It is at this point in the argument that readers might be reminded of the quotation from Professor J. B. S. Haldane, the geneticist:

> If the co-operation of some thousands of millions of cells in our brain can produce our consciousness, the idea becomes vastly more plausible that the co-operation of humanity, or some sections of it, may determine . . . a Great Being.

Teilhard now suggests that we begin to see why fervour and impotence accompany every egoistic solution of life.

Egoism may feel right but its mistake is in confusing individuality with personality. In trying to separate itself as much as possible from others, the element individualizes itself—and in doing so it becomes retrograde and seeks to drag the world backwards. In fact it diminishes itself and loses itself. To be fully ourselves it is in the opposite direction, in the direction of convergence with all the rest, that we must advance.

The peak of ourselves, the acme of our originality, is not our individuality but our person: and according to the evolutionary structure of the world we commonly find our person by uniting together. There is no mind without synthesis. The element becomes personal only when it universalizes itself.

If we want to give effective help to the progress of evolution in ourselves, the energies we must identify, harness and develop before all others are those of a social nature.

Which brings us to the problem of love as energy.

Love as Energy

Love in its biological sense is not peculiar to man. It is a general property of life and as such embraces all the forms successively adopted by organized matter. In mammals, for example, it is easily recognized in its different forms: sexual passion, parental instinct, social solidarity, etc. As we go down the scale of life the analogies are more obscure, until they become so faint as to be almost imperceptible.

But here again Teilhard brings in the Within of things, without which there would be no real internal propensity to unite. We should, therefore, in his opinion, assume its presence in everything that exists. In fact, when we take into account the ever flowing together of consciousness, we see that it is not lacking anywhere. Plato felt this and immortalized the idea in his Dialogues.

To perceive cosmic energy at its fount we must go down into the internal or radial zone.

Love alone is capable of uniting living beings in such a way as to complete and fulfil them, for it alone takes them and joins them by what is deepest in themselves.

When do lovers come into the most complete possession of themselves if not when they say that they are lost in each other? If love can achieve this in a "couple" or a "team", why should it not be able to achieve it one day throughout the world.

Such an idea of the synthesis of individuals and peoples is said to be Utopian. But it is in fact biologically necessary to

84

the future of the world. All we may well need is to imagine our power of loving developing until it embraces all men and all the earth.

If you claim, as you well may, that a universal love is impossible, how do we account for the irresistible instinct in our hearts which leads towards unity when our passions are stirred? A sense of the universe, a sense of the "all", the nostalgia which seizes us when confronted by nature, beauty, music—these seem to be an expression and awareness of a Great Presence.

We are inclined to think that we have exhausted the various material forms of love with a man's love for his wife, his children, his friends, and to a certain extent his country. Yet the most fundamental form of this passion is missing from this list—namely that of the Whole, the cosmic affinity and cosmic sense. A universal love is not only psychologically possible, it is the only complete and final way in which we are able to love.

This brings Teilhard to the point of asking how to overcome the appearance all around us of mounting hatred? He says we should learn to overcome this anti-personalist complex which paralyses us and to make up our minds to accept the possibility and the reality of some source of love, and object of love, at the summit of the world.

So long as collectivity absorbs, or appears to absorb, the person, it kills the love that is trying to come to birth. As such, collectivity is essentially unlovable. That is where philanthropic systems break down. Common sense is right. It is impossible to give oneself to an anonymous number. But if the universe ahead of us assumes a face and a heart, and so to speak personifies itself, not of course by becoming a person but by charging itself at the heart of its development with a focus of personal energies, then in the atmosphere created by this focus the elemental attraction will immediately blossom.

The discoveries of the last hundred years with their unifying perspectives have brought a new and decisive impetus to our sense of the world, our sense of the earth and to our human sense.

It is necessary in Teilhard's opinion for us to extend our science to its furthest limits, to recognize and accept not only some vague future existence (and our new concept of space-time emphasizes this) but also the radiation as a present reality of this mysterious centre of centres. Teilhard has called this centre of centres the Omega Point, by which he probably means the ultimate essence of the universe.

The Attributes of Teilhard's Omega Point

In Teilhard's view modern thought is at last getting used to the idea of the creative value of synthesis in evolution. There is more in the molecule than in the atom, more in the cell than in the molecule, more in society than in the individual and more to mathematical construction than calculations and theorems.

We are now inclined to admit that at each further degree of combination, something which cannot be reduced to isolated elements emerges in the new order of things.

With this admission, consciousness, life and thought are on the threshold of acquiring a right to existence in terms of science. However, science is still far from recognizing that this something has independence and solidity. In spite of a half-hearted conversion to "spiritual" views, it is still towards matter as being infinitely diluted that scientists look, when they seek the eternal and the Great Stability.

In further considering the attributes of Omega, Teilhard believes that it is necessary to rid ourselves of a restricted outlook. He advances two reasons for doing so—one of love and one of survival.

In the case of love, Teilhard believes that since this dies in contact with the impersonal and anonymous, and equally perishes from remoteness in space and time, Omega must be in some manner loving and lovable at this very moment. It must be present throughout the stuff of the universe.

In the case of survival, Teilhard asks: what is the use of detecting a focus of this kind in the van of evolution if that focus will one day disintegrate? To satisfy our ultimate requirements Omega must be independent of the collapse of the forces with which evolution is woven. It must be autonomous.

To be such, it would not be enough to emerge from the rise of consciousness. It would have to have already emerged. If Omega by its very nature did not escape from the time and space it gathers together, it would not be what it needs to be.

To autonomy we therefore must add ever-presentness, actuality. And to these two we can add irreversibility: Omega would not be Omega if the process could start going backwards. Finally, we add transcendence: Omega must transcend everything.

Contrary to the views still held by the physicists, Teilhard says, the Great Stability is not at the bottom in the infinitely diluted, but at the top in the ultra-synthesized. It is entirely in its tangential envelope that the world goes on dissipating itself into matter. In its radial nucleus it finds its shape and natural consistency in gravitating against the tide of probability towards a divine focus of mind which draws it onwards.

Once reflection came into play, the elements could begin to react to the personalizing action of the centre of centres.

From the grains of thought forming the veritable and indestructible atoms of the stuff of the universe, this universe goes on building itself above our heads in the inverse direction of matter.

15

THE ULTIMATE EARTH

WE HAVE SEEN that without the involution of matter upon itself, without the enclosed chemistry of molecules, cells and phyletic branches, there would never have been any biosphere or noosphere. In their advent and their development, life and thought are not only accidentally, but also structurally, bound up with the contours and destiny of the terrestrial mass.

No one would dare to picture what the noosphere will be like in its final guise, no one, that is, who has glimpsed the incredible potential of unexpectedness accumulated in the spirit of the earth.

The end of the world defies imagination. We may nonetheless to some extent foresee the significance and circumscribe the forms of the ultimate earth.

Views To Be Set Aside

Whenever the end of the world is mentioned, the first idea that leaps to our minds is always one of catastrophe. Generally we think of a sidereal cataclysm. There are so many stars hurtling around and brushing past, there are those exploding worlds on the horizon. Surely, we say, our turn will come sooner or later and we shall be stricken or killed, or at least have to face slow death in our terrestrial prison.

Since physics has discovered that all energy runs down,

we seem to feel the world getting a little cooler every day. This, however, has been partially compensated for by the discovery of radio-activity which has happily intervened to delay the imminent cooling.

The astronomers are now in a position to predict that, if all goes as it should, we have at any rate several hundred million years ahead of us. Nevertheless, we are even more threatened by internal dangers at the level of both the biosphere and the noosphere. Onslaughts of microbes, sterility, war, revolution—there are so many ways of our world coming to an end.

Yet despite numerous forecasts of manifold disasters, Teilhard feels entitled to say that we have nothing to fear in so far as such disasters imply the idea of premature accident or failure. Pessimists, he says, are inclined to take individual conditions and apply them to life as a whole. Because accident, disease and decrepitude spell death to some men, the same does not apply to mankind as a whole.

We must not forget that, since the birth of thought, man has been the leading shoot of the tree of life. The hopes for the future of the noosphere are concentrated exclusively upon him as such. The world in its present state would be unintelligible unless we supposed an implicit complicity between the infinite and the infinitesimal to warm, nourish and sustain, to the very end, the consciousness that has emerged between the two.

It is on this complicity that we must depend. Man is irreplaceable. Therefore, however improbable it may seem, he must reach his goal—infallibly.

What we expect is an ultimate progression inevitably coming at its biologically appointed hour. It is in this direction we should look if we want a forward glimpse of the end of the world.

The Approaches

Compared with past zoological layers, whose average duration is at least of the order of eighty million years, mankind is so young that it could almost be called newborn. On the other hand, to judge from the rapid development of thought in the short period of a few dozen centuries, this youth bears the indications and promise of an entirely new biological cycle.

In all probability there now stretches an immense period characterized by a speeding-up and a flowering of the forces of evolution along the line of the human shoot. Teilhard next depicts the form and lines along which progress could develop during this period.

First in a collective form. He is doubtful whether the physical form of man will undergo any material transformation. He considers that evolution in future will be occupied in a richer and more complex domain, namely the realms of the mind and in particular that of the communal mind.

He visualizes three principal lines of advance: the increased organization of research; the development of concentrated research on the subject of man; and the conjunction of science and religion.

A. *The organization of research*

In Teilhard's opinion too much effort is devoted to the business, first, of killing each other, and secondly, to the increase of industrial production and consumer production. Not enough effort, he says, is being put into the pursuit of truth. Less money is provided for pure research all over the world than is laid out on one capital ship.

He does not think that we are fully conscious of or in control of the new powers that have been unleashed. We see in science only a new means of providing more easily the same old things. But the time will come when we shall be forced to admit that science is not an accessory occupation but an essential activity.

More experimentation should be undertaken. Giant telescopes and atom smashers should attract more money and excite more enthusiasm than equipping any armed forces.

It is Teilhard's hope that it will not be long before the noosphere finds its "eyes".

B. *Research on man*

If life has been able to advance, it is because by ceaseless groping it has successively found the points of least resistance at which reality yielded to its thrust. Similarly, if research is to progress tomorrow, it will be largely by localizing the sensitive zones whose conquest will afford us easy mastery of all the rest.

Man will perceive at last that man as the object of knowledge is the key to the whole science of nature—the provider of everything we can know.

Up to the present, science has fought shy of looking man in the face, as it were, and has circled round the whole human being seeing him as an object without daring to investigate the whole truth.

Teilhard believes wholeheartedly in the pre-eminent significance of man in nature. He feels that there is still no certainty whether the physicist is dealing with pure energy or with manifestations of thought. Biology, too, finds itself forced to acknowledge thinking beings as the present ultimate form of evolution.

In a striking phrase he says:

We find man at the bottom, man at the top, and, above

all, man at the centre ... We shall have to get to grips with him sooner or later.

To decipher man is essentially to try and find out how the world was made and how it ought to go on making itself. It means profound study of the past and of origins. It means constructive experiment on continually renewed objects.

What is involved, primarily, is the care and improvement of the human body, the health and strength of the organism. In the course of the coming centuries it is indispensable that an advanced method of human eugenics, of a standard worthy of our personalities, should be discovered and developed.

In society in future there should be a harmonious reconciliation between what is free and what is planned. Points involved would include: the equitable distribution of the world's resources, the maximum use of the powers set free by mechanization, the physiology of nations and races, international collaboration in economics, politics and population trends.

C. *Science and Religion*

In the course of such a future as has been outlined above, science by being led to concentrate on man will find itself increasingly face to face with religion.

We have heard much in modern times of the conflict between science and religion. Indeed, at one moment it seemed a foregone conclusion that the former was destined to take the place of the latter. As the tension is prolonged, the conflict visibly seems to need to be resolved not in terms of elimination, nor in terms of duality, but in terms of synthesis.

In two centuries of passionate struggles, neither side has succeeded in discrediting its adversary. Neither can develop naturally without the other, for the same life animates both.

When in the universe of movement to which we have just awakened we look at the temporal and spatial elements diverging and amplifying themselves, we are perhaps engaging in pure science; but when we turn towards the summit, towards the totality and the future, we cannot help introducing religion.

Religion and science are two conjugated forces or phases of one and the same complete act of knowledge—the only one which can embrace both the past and the future of evolution.

In the mutual dependence of these two supposedly opposed powers, science and religion, the human spirit is destined to extend itself to the maximum.

The Ultimate

Mankind has enormous possibilities before it. At the point that Man began to have knowledge of himself he entered into an entirely new field—thanks not only to the great advances in the development of individual thought, but also to the prodigious power of the communal thought of human beings as a whole. Teilhard suggests that this layer of thought—this noosphere—is destined, by reflecting upon itself, to form a new and final organized unity. Atoms, molecules, cells and human personalities have all been examples of such a process. The noosphere will rise up in us and through us unceasingly and eventually will centre itself in a single point of ever increasing concentration.

Eventually Mankind will have reached a true perception of the ultimate essence of the Universe. *This will be the final fulfilment of the spirit of the Earth.*

Appendix

TEILHARD ON EVIL

IT SEEMS APPROPRIATE to point out that Teilhard adds
to the main text of *The Phenomenon of Man* certain remarks
on the place and part of evil in a world in evolution:

> Throughout the long discussions we have been through, one
> point may perhaps have intrigued or even shocked the reader.
> Nowhere have pain or wrong been spoken of. Does that mean
> that ... evil and its problem have faded away and no longer
> count in the structure of the world? If that were so, the picture
> of the universe here presented might seem over-simplified or
> even faked.

Teilhard's answer is:

> My aim has been limited to bringing out the *positive essence*
> of the biological process of hominization ... What good
> would it have done to have drawn attention to the shadows on
> the landscape, or to stress the depths of the abysses between
> the peaks? ... I have assumed that what I have omitted could
> nevertheless be seen. And it would be a complete misunder-
> standing to interpret the view here suggested as a sort of human
> idyll rather than as the cosmic drama I have attempted to
> present. True, evil has not hitherto been mentioned, at least
> explicitly. But on the other hand surely it inevitably seeps out
> through every nook and cranny, through every joint and
> sinew of the system.

First: *evil of disorder and failure.* Right up to the reflective

95

zones we have seen the world proceeding gropingly and by chance. How many failures must there have been for one success? There is the physical disarray on the material level; then there is suffering, inherent in the flesh; and on a still higher level, the torture of spirit in self-analysis.

Second: *evil of decomposition.* This is one form of the first type. Sickness and corruption invariably result from some unhappy chance. It is an aggravated type, it must be added, since for living creatures death is the regular, inescapable method of replacing one individual by another. Death is the indispensable lever for the rise of life.

Third: *evil of solitude and anxiety.* This is the great anxiety (peculiar to man) of consciousness wakening to reflection in a dark universe which light takes centuries and centuries to reach, a universe we have not yet succeeded in understanding either in itself or in its demands on us.

Last: *evil of growth.* The least tragic perhaps, because it exalts us, though it is nonetheless real. We find it in the pangs of childbirth and in the mysterious law which makes all progress towards increased unity express itself in terms of labour.

Indeed, looking at the march of the world from the point of view not of its progress but of the risks and efforts it demands, we see that evil appears, as abundantly as anyone could care to conceive it, in the path of evolution through the very structure of the system.

Pain and wrong, tears and blood: they are the by-products (often precious, and re-usable) created by the world of thought in its progress. This, in the final analysis, is what the sight of the world in movement reveals to our observation and reflection.

Even in the view of the biologist, the human epic resembles nothing so much as the way of the Cross.

Author's Footnote

FINDING THE SIGNS

IT IS MORE than a little difficult to say where, at this point, the signs of such a fulfilment of the spirit of the Earth are to be found. "Peace among Nations" has been a dream of the idealists long before and since the coming of Christ. What is the evidence for such an ideal at the tail-end of the twentieth century? Precious little. Besides two world wars, this century has seen the creation of two international peace organizations, first the League of Nations and subsequently the United Nations, but neither has to date realized initial expectations. One might suppose that religion too would provide some positive evidence, but, here too, not only has there been hostility between the different religions of the world, but also internal enmities within them.

Teilhard's view is that, as the world progresses, these differences of view will slowly and surely modify, and that in due time we shall approach nearer and nearer to his picture of a world more or less at one with itself. It must be frankly admitted that this has been a dream throughout the ages, a dream that seldom if ever has approached the stage of practical realization. Nevertheless, Teilhard's remarkable work does provide a glimmer of hope that we may be nearer to this goal.

A full understanding of his ever recurring theme, namely, the Within and Without of all things, provides the key to the essence of his work. In view of the sorry state of disunity in which the world finds itself today it would appear to be well worth while examining the whole of Teilhard's thesis outlined in his *Phenomenon of Man*. Anyone who is prepared to do this would understand more clearly and more fully the vital character of his vision.

Teilhard throws a bridge across the age-long chasm between science and religion, but this little book highlights his insistence that it is the man-made barriers that stand in the way of mankind's final fulfilment of the spirit of the Earth.

M. K.

INDEX

INDEX

Multiplication, 58
Mutation, 67

Neanderthal man, 53, 54, 55
Neolithic Age, 58–63, 75
Nile, 62
Noogenesis, 48
Noosphere, 48, 49, 60, 64, 77, 78, 82, 89, 90, 92, 94
Nucleus, 1, 2

Omega point, 86, 87
Omnivores, 33
Organic compounds, 11, 12
Orthogenesis, 25
Oxides, 11

Pacific, 59
Palaeontology, 30, 31, 34, 40, 55
Peduncle, 51
Peking man, *see under* Sinanthropus
Permian era, 34
Personalization, personality, 40, 48, 81, 83, 87, 94
Phylum, 27, 28, 29, 30, 31, 47, 51, 55, 56, 60, 61, 73, 76, 89
Physics, 6, 15, 43, 89, 92
Pithecanthropus, 53
Placentals, 33, 34
Planetization, 73
Plants, 6, 35
Plato, 84
Pliocene era, 51
Post-Neolithic man, 61
Pre-history, 60
 -hominids, 54, 55
 -Jurassic period, 34
 -life, 12, 13, 14, 16, 20
Primates, 40, 41
Profusion, 26
Protoplasm, 16, 18, 20
Pseudo-Neanderthal man, 55
Psyche, psychism, 40, 44, 45, 47

Quadrupeds, 34, 41
Quantum, 22
Quaternary era, 55, 56

Racialism, 71
Radial energy, *see* Energy
Radio, 72
Radio-activity, 90

Reflection, 44, 45, 46, 48, 49, 50, 51, 52, 59, 68
 threshold of, 45, 48, 50, 51, 65
Religion, 91, 93, 94
Reproduction, 23, 24, 47
Reptiles, 34, 38
Research, 91, 92
Rome, 62

Science, 1, 4, 7, 43, 44, 50, 76, 77, 86, 91, 92, 93, 94
Sex, 47, 84
Sinanthropus, 50, 53, 54
Socialization, 29, 58
South Seas, 62
Space, 3, 18, 65, 68, 81, 87
Space-time, 65, 66, 68, 76, 81, 82, 86
Spirit, 72, 74, 79, 94
Stars, 3, 4, 18, 67, 68, 81, 89
Stratosphere, 11, 49
Super-consciousness, 76, 77, 78
 -human, 74
 -life, 70
 -man, 71
 -physical, 77
Survival, 70
Symbiosis, 74
Synthesis, 13, 61, 76, 82, 83, 84, 86, 93

Tangential energy, *see* Energy
Tertiary era, 48
Thermodynamics, 13
Thought, 9, 42, 43–9, 54, 69, 70, 72, 74, 78, 86, 88, 89, 90, 91, 92, 94
Time, 21, 65, 68, 87
Tree of life, *see* Life, tree of

Ultimate, 94
Unity, 2, 26, 83, 84, 85
Universe, 1–4, 18, 76, 82, 85, 94
 stuff of the, 2, 4, 12, 17, 20, 23, 65, 88
Upper Palaeolithic man, 55, 56

Vertebrates, 34, 35

Within of things, 5–10, 12, 17, 23, 37, 39, 44, 46, 72, 74, 84
Without of things, 5, 6, 12, 23, 37, 44, 46, 74

Yellow River, 62

Zoology, 6, 16, 47, 67, 73, 91